THIS IS
geography

JOHN WIDDOWSON

2

HODDER
EDUCATION
AN HACHETTE UK COMPANY

Orders: please contact Bookpoint Ltd, 130 Milton Park, Abingdon, Oxon OX14 4SB. Telephone: +44 (0)1235 827720. Fax: +44 (0)1235 400454. Lines are open 9.00–5.00, Monday to Saturday, with a 24-hour message answering service. Visit our website at www.hoddereducation.co.uk

© John Widdowson 2008
First published in 2008 by Hodder Education,
an Hachette Company UK
338 Euston Road
London NW1 3BH

Impression number   5   4
Year   2013

Illustrations by Oxford Illustrators, Stephanie Strickland, Richard Duszczak and Mark Duffin
Layouts by Lorraine Inglis Design
Typeset in Meridien 12/14pt
Printed and bound in Dubai

A catalogue record for this title is available from the British Library

ISBN: 978 0 340 90742 9

Teacher's Resource Book including CD-ROM
978 0 340 90745 0

Contents

Key features of *This is Geography*

Before you start *This is Geography*, here is a quick guide to help you find your way around. Book 2 is split into eight units, covering eight enquiries. In each one you will find the following features:

The opening spread

The unit title

Key concepts covered in this unit.

The big enquiry question – the main question you will focus on through the unit.

2 Landscape detective

What can you learn about a place by investigating rocks?

KEY CONCEPTS
- Place
- Physical processes
- Space

WHAT CAN YOU LEARN ABOUT A PLACE BY INVESTIGATING ROCKS?

You'll be surprised at how many rocks you can find in your street! Look out for four types of rock: igneous, sedimentary, metamorphic or made by people.

coming up...
Have you ever wondered why every landscape looks so different? The answer lies under your feet!

The rock beneath the ground makes each landscape unique. Even in cities, many buildings are made from rocks. If we dig a bit deeper, rocks also give us clues about how the landscape was formed.

You are a landscape detective. You will look for clues about how the landscape is formed. You will start on a trail around your local town, and then end up in the Yorkshire Dales.

your final task...
At the end of the unit, you will plan a trail around the Yorkshire Dales. This one will be a bit more adventurous!

Rock made by people is used for many building jobs. For example, bricks are made by heating clay. Concrete, glass and ceramic tiles are man-made too.

IGNEOUS ROCK is made of crystals. You can see the crystal pattern in most igneous rocks. This is granite, a common igneous rock often used for kerbstones and shop fronts.

SEDIMENTARY ROCK is made from sediment – bits of clay, sand, stones or shells, all stuck together. Limestone is a common sedimentary rock used for building. This is travertine, a type of limestone, used on the front of McDonald's restaurants.

METAMORPHIC ROCK is more difficult to recognise. It is formed from igneous or sedimentary rock that has been changed by heat and pressure. Slate is a metamorphic rock that was once clay. It is used for roof tiles.

LOCAL INVESTIGATION

investigate...
1 Follow a trail around your local town centre. Your teacher may give you a trail to follow or you can make up your own trail on a map.
  a) How many different kinds of rock can you find? Draw a sketch (or take a photo) and write down anything you notice about each rock you find.
  b) Mark where you find each rock on a map.

aim high...
2 When you come back to the classroom you can try to identify the names of each rock. You could use rock identification tables on the 'About:Geology' website at: www.geology.about.com/library/bl/blrockident_tables.htm to help you. Click on any rock in the table to see a photo. Don't worry if you can't find the exact rock. Being a landscape detective isn't always easy!

Coming up – tells you what you are going to do through the rest of the unit.

Your final task – what you will be doing at the end of the enquiry to bring all your work together.

Key words – in SMALL CAPITALS. They are explained in the Glossary on pages 137–138.

Through the unit

The enquiry question – repeated on every spread so you won't forget it!

Activity – tasks that will help you to build on your enquiry, step by step.

Aim high – a challenge, and not just for the clever ones! It is a task that will help you to take your geography that little bit further.

The final spread

The enquiry question – again! You should know it by now.

Your final task – this is what all your work has been leading to. It is your chance to show what you know and what you can do. You may be asked to:
- draw a plan
- make a wall display
- write a story
- design a poster
- make a PowerPoint presentation.

1 Welcome to Earth Village

Can the Earth cope with any more people?

The world has about six and three-quarters billion people. That is **6,750,000,000**!

Every year the POPULATION grows by another 90 million – twice the population of England. Can the Earth cope with any more people?

▮▮ coming up...

It's impossible to imagine so many people. But, picture the world as a village with just one hundred people. Let's call it Earth Village. This drawing shows what is happening in Earth Village. The same things happen in the real world too. You will find out more about Earth Village. It will help you to understand more about the world's population.

▮ your final task...

At the end of the unit you will decide if the Earth can cope with any more people. Are you an optimist or a pessimist?

Some of the ideas in this unit were inspired by a book called, *If the world were a village* by David J Smith. If you enjoy this unit, you will enjoy reading the book.

One hundred people live in the village, but the population is growing. By next year there will be 102.

People in the village pollute the environment. They burn fires, drive cars and throw rubbish away.

Some areas are empty.

activity...

1 Look carefully at Earth Village.
   a) Which part of the village is most like the area that you live?
   b) How does this area differ from other parts of Earth Village?

2 What does the drawing tell you about:
   a) the population of Earth Village? Mention three things.
   b) the problems that Earth Village faces? Mention three more things.

Some parts of the village are crowded.

SHOPPING CENTRE

It's a boy!

...e dies and the ...on goes down.

A baby is born, so the population goes up.

People of all ages live in the village. Most of them are young.

The village depends on natural resources. People get water from the river, food from the land and sea, and fuel from below ground. Some resources are running low.

...e village are ... Some people ...ers are poor.

discuss...

3 If you lived in Earth Village would you be optimistic or pessimistic about the future? Think of reasons. You will come back to this question at the end of the unit.

→ Population explosion

10,000 BC
It's not much fun with just the two of us.
We need to multiply!

0 AD

1500 AD
That's better! More people to meet!
And more ideas!

1800 AD
The village is getting more crowded!
Let's spread out!

This is how the population of Earth Village has grown.

Earth Village is more crowded now than it's ever been before. It's the same for the real world. Look at graph **A**. It took the whole of human history up to 1800, for the world's population to reach one billion. By 1960, it had grown to three billion, and it doubled again to six billion by 1999. Today, the world is gaining one billion more people every eleven years. No wonder we call it the POPULATION EXPLOSION!

activity...

1 Look at graph **A**.
   a) Find out the years when the world's population reached,
      i) 500 million ($\frac{1}{2}$ billion)   ii) 1 billion   iii) 2 billion
      iv) 4 billion
   b) In each case, work out how long it took for the population to double. What do you notice?
   c) Predict when the world's population is likely to reach 8 billion.
2 Label these events on a large copy of the graph using the dates. For each event, draw an arrow to the correct point on the line. One is done for you.

10,000 BC	People start to farm. Before, they ate wild plants and animals.
5000 BC	The first towns and cities grow in the Middle East.
0 AD	Rome has a million people. The Roman Empire expands.
1500 AD	Europe is getting crowded. People start going to other parts of the world.
1750 AD	The Industrial Revolution begins and cities grow.
1850 AD	People learn to control some diseases.
2000 AD	Over half of us now live in cities.

A World population growth. Note that the time scale on the bottom of the graph changes from BC to AD.

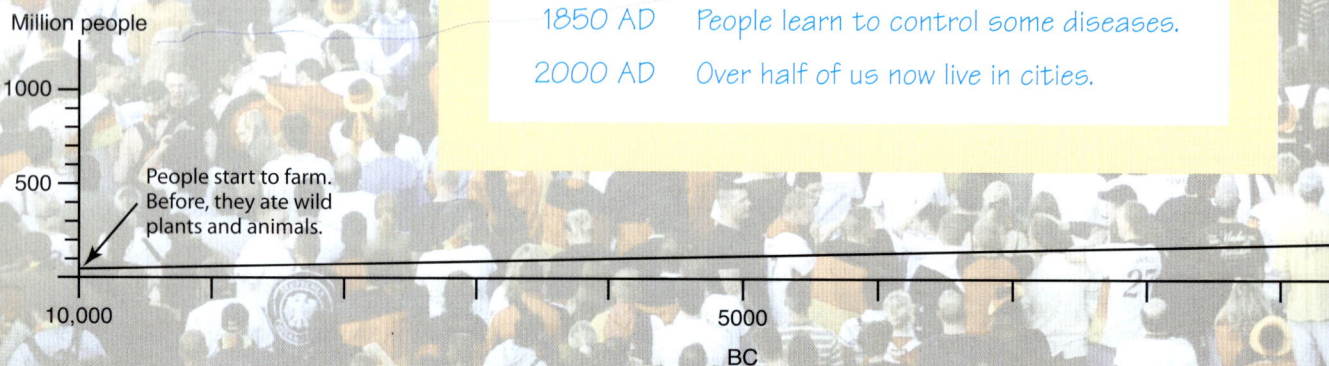

Million people

1000 —

500 —

People start to farm. Before, they ate wild plants and animals.

10,000

5000

BC

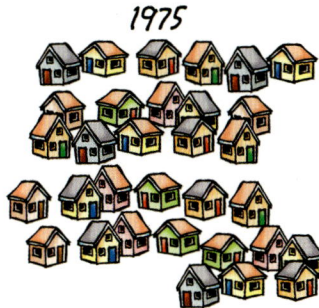

1900

1975

2005

There's too many people, and not enough resources!

There's enough if we share!

Billion people

aim high...

3 a) Find out what the world's population is today. You can find out from a population clock on the Internet. Here is one you can try: www.populationpress.org

b) Compare the population now with the prediction that you made in activity **1**. Does it look like your prediction will come true? If not, revise your prediction taking the latest population into account.

7

6

5

4

3

2

1

500

1000

1500

2000

AD

→ Births and deaths

Population grows because the number of babies being born (the BIRTH RATE) is greater than the number of people dying (the DEATH RATE). The POPULATION GROWTH RATE is the difference between the birth rate and death rate.

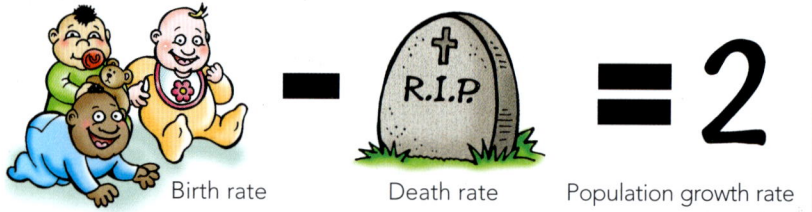

Birth rate   Death rate   Population growth rate

The population of Earth Village is growing. On average, three babies are born and one person dies each year, so the population grows by two.

Food shortage

Children working for family

No more work. We can relax!

Old age pensions

War

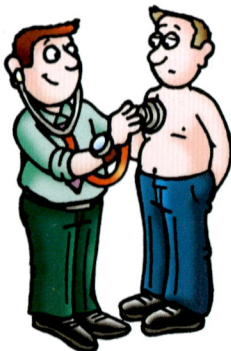

Hospitals and doctors

Education

Birth control

Good harvest

In the real world, population growth varies from one country to another. In rich countries (like the UK) there is not much difference between birth and death rates, so the population growth is slow. In poor countries (like Kenya) birth rates are much higher than death rates, and so the population grows faster. The illustrations in **B** below show the factors that affect birth rates and death rates in different countries.

B Factors that affect birth rates and death rates

Clean water supply

Traditional to have large families

Disease

High infant mortality (children die as infants)

activity...

1 Look at the factors in **B**. Think how they could affect either the birth rate or the death rate.

  a) Classify the factors into four groups:
  - factors that lead to high death rates
  - factors that lead to low death rates
  - factors that lead to high birth rates
  - factors that lead to low birth rates.

  List the factors under the four headings. For example, 'children working' goes under 'Factors that lead to high birth rates'. You should have three factors under each heading.

  b) Choose one factor from each group. Write a sentence to explain how it affects the birth rate or death rate. For example, Children working leads to a high birth rate because poor families have more children to support them.

2 Look at the table below, comparing birth rates and death rates in the UK and Kenya.

	UK	Kenya
Birth rate (per 1000 people)	11	39
Death rate (per 1000 people)	10	11
Population growth rate (per 1000 people)	?	?

  a) Copy the table. Work out the population growth rate for the UK and Kenya.
  b) Write three sentences to compare the birth rate, death rate and population growth rate in the UK and Kenya.

The birth rate in Kenya is roughly _____ times _____ than in the UK.

The death rate in Kenya is _____.

Population growth rate in Kenya is _____.

aim high...

3 Suggest reasons why population growth in Kenya is higher than in the UK. (The factors in **B** should give you some ideas.)

9

➤ Young and old

How long do you expect to live? Your answer could depend on where you live. LIFE EXPECTANCY is the average number of years that a person can expect to live. Your life expectancy will vary depending on the country you live in.

On average, people in rich countries (like the UK) have a greater life expectancy than people in poor countries (like Kenya). You can see the differences when you look at the population pyramids (**C** and **D**) for the two countries. A population pyramid shows the numbers of people in each age group.

Population pyramids are not as complicated as they look. Really, they are bar graphs with two sides. On the left side are males, and on the right, females. The youngest age group is at the bottom and the oldest is at the top. There are usually more young people than old people, so the graph is shaped like a triangle. That's why they are called pyramids.

> It's been a good life. Bu[t] I'm looking forward to a long retirement!

In Earth Village the average life expectancy is 63. However, in some parts of Earth Village the life expectancy is higher, and people live longer. In other parts it is lower.

C Population pyramid for Kenya

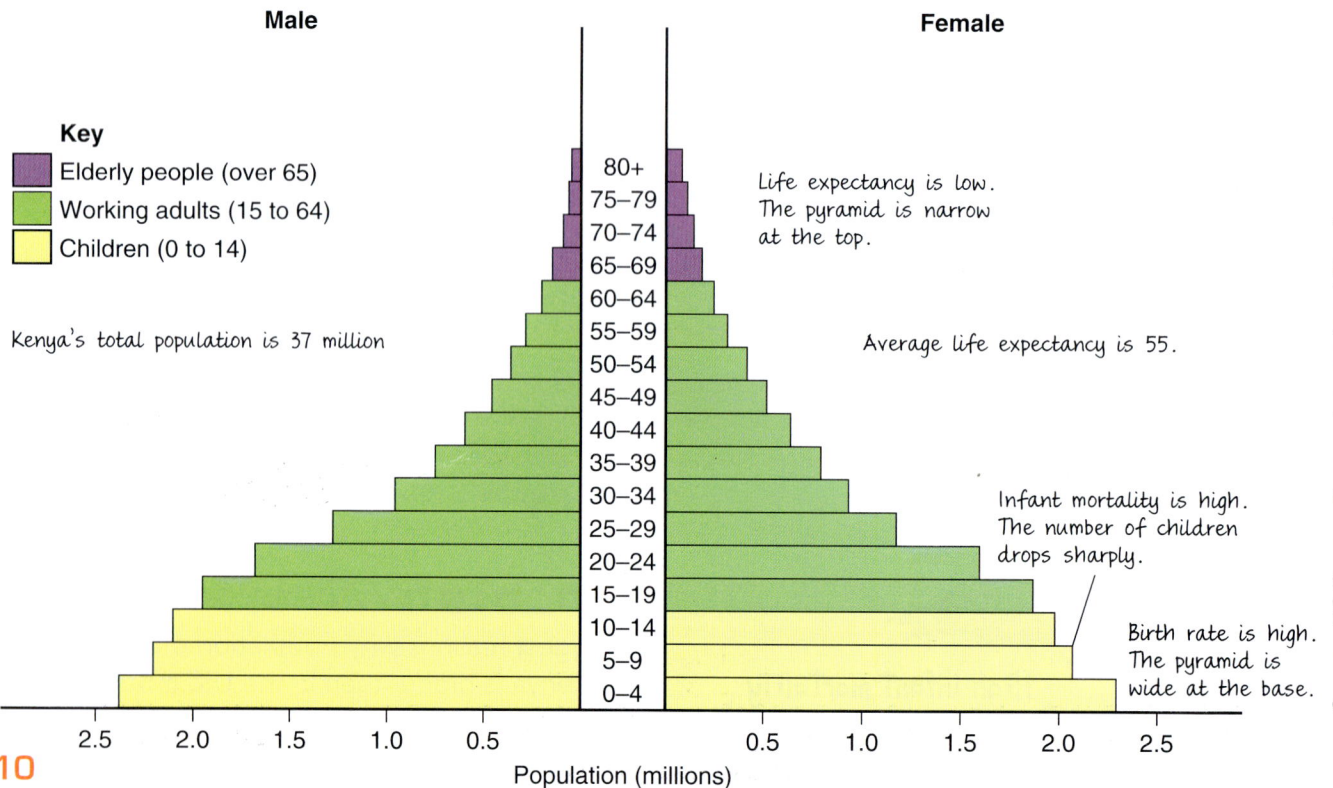

Male

Female

Key
- Elderly people (over 65)
- Working adults (15 to 64)
- Children (0 to 14)

Kenya's total population is 37 million

Life expectancy is low. The pyramid is narrow at the top.

Average life expectancy is 55.

Infant mortality is high. The number of children drops sharply.

Birth rate is high. The pyramid is wide at the base.

Age groups (top to bottom): 80+, 75–79, 70–74, 65–69, 60–64, 55–59, 50–54, 45–49, 40–44, 35–39, 30–34, 25–29, 20–24, 15–19, 10–14, 5–9, 0–4

Population (millions): 2.5 2.0 1.5 1.0 0.5 | 0.5 1.0 1.5 2.0 2.5

Population (millions)

Source: US Census Bureau, International Database

activity...

1 Look at the population pyramids, **C** and **D**. Pyramid **C** shows the population of Kenya. Read the labels to help you to interpret the pyramid.

a) Label a copy of pyramid **D** in the same way to describe the population in the UK. Label the correct parts of your pyramid with the descriptions below.

> Life expectancy is high
>
> Birth rate is low
>
> Infant mortality is low
>
> Women have greater life expectancy than men

b) Write sentences using each label to explain what the pyramid shows. For example, Life expectancy is high. The pyramid is wide at the top.

aim high...

2 You are going to use a computer model to find out how Kenya's population is likely to change in future. You can do the same for the UK.

a) Go to International Database at the US Census Bureau website: www.census.gov/ipc/www/idb/pyramids.html. Select 'Kenya'. Under 'Type of output' choose 'Dynamic', then 'Submit'. The dynamic pyramid shows how the population of Kenya is expected to change up to 2050.

b) Print a copy of the population pyramid in 2050. Label the changes on the pyramid since 2008 and stick it in your book.

Then, you can do the same activity for the UK.

D Population pyramid for the UK

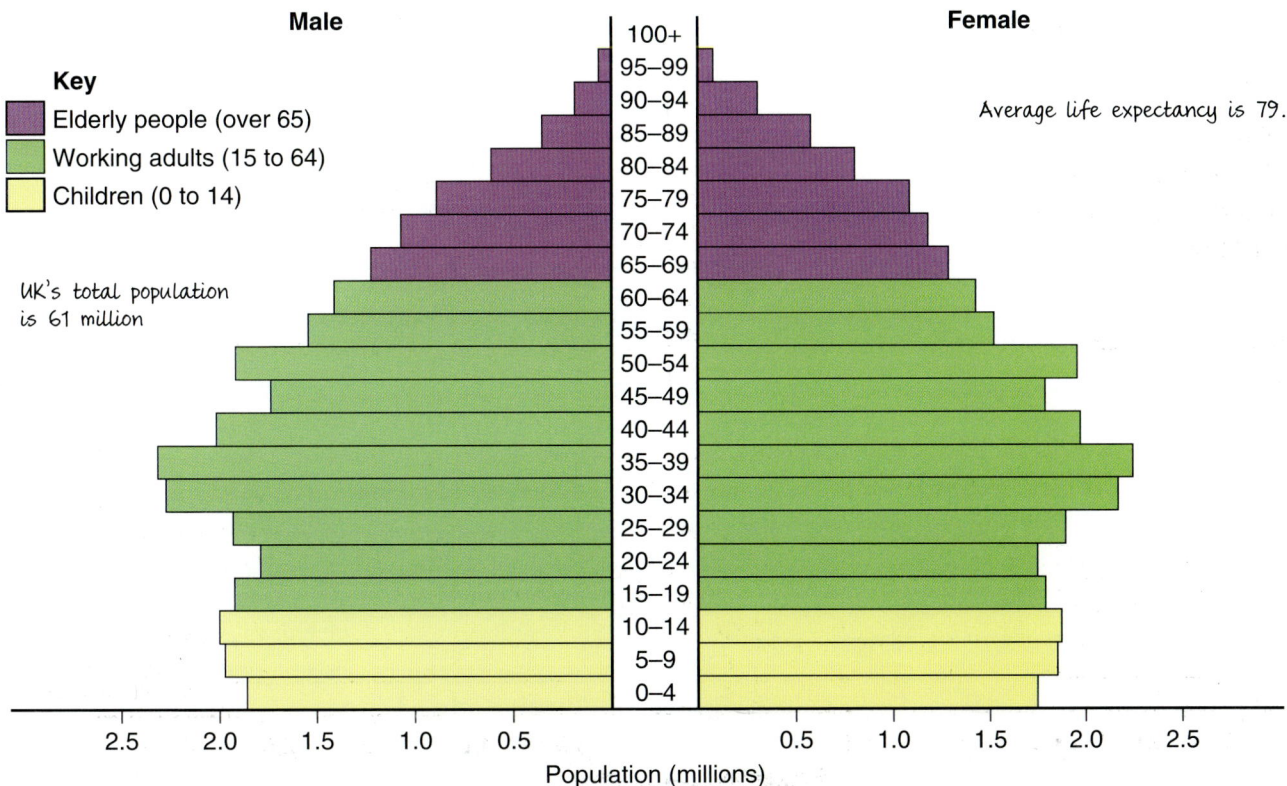

Male / Female

Key
- Elderly people (over 65)
- Working adults (15 to 64)
- Children (0 to 14)

UK's total population is 61 million

Average life expectancy is 79.

Age groups (top to bottom): 100+, 95–99, 90–94, 85–89, 80–84, 75–79, 70–74, 65–69, 60–64, 55–59, 50–54, 45–49, 40–44, 35–39, 30–34, 25–29, 20–24, 15–19, 10–14, 5–9, 0–4

Population (millions): 2.5 2.0 1.5 1.0 0.5 | 0.5 1.0 1.5 2.0 2.5

Source: US Census Bureau, International Database

→ Mapping population

Imagine trying to make a map to show where everyone in the world lives. It would not be easy! In Earth Village there are 100 people and, like the real world, they are not evenly spread around.

Each person in Earth Village is represented by a dot on map **E**. This map shows the way that people are spread, or the POPULATION DISTRIBUTION. Areas where most people live are densely populated. Areas where few people live are sparsely populated.

E Population distribution in Earth Village

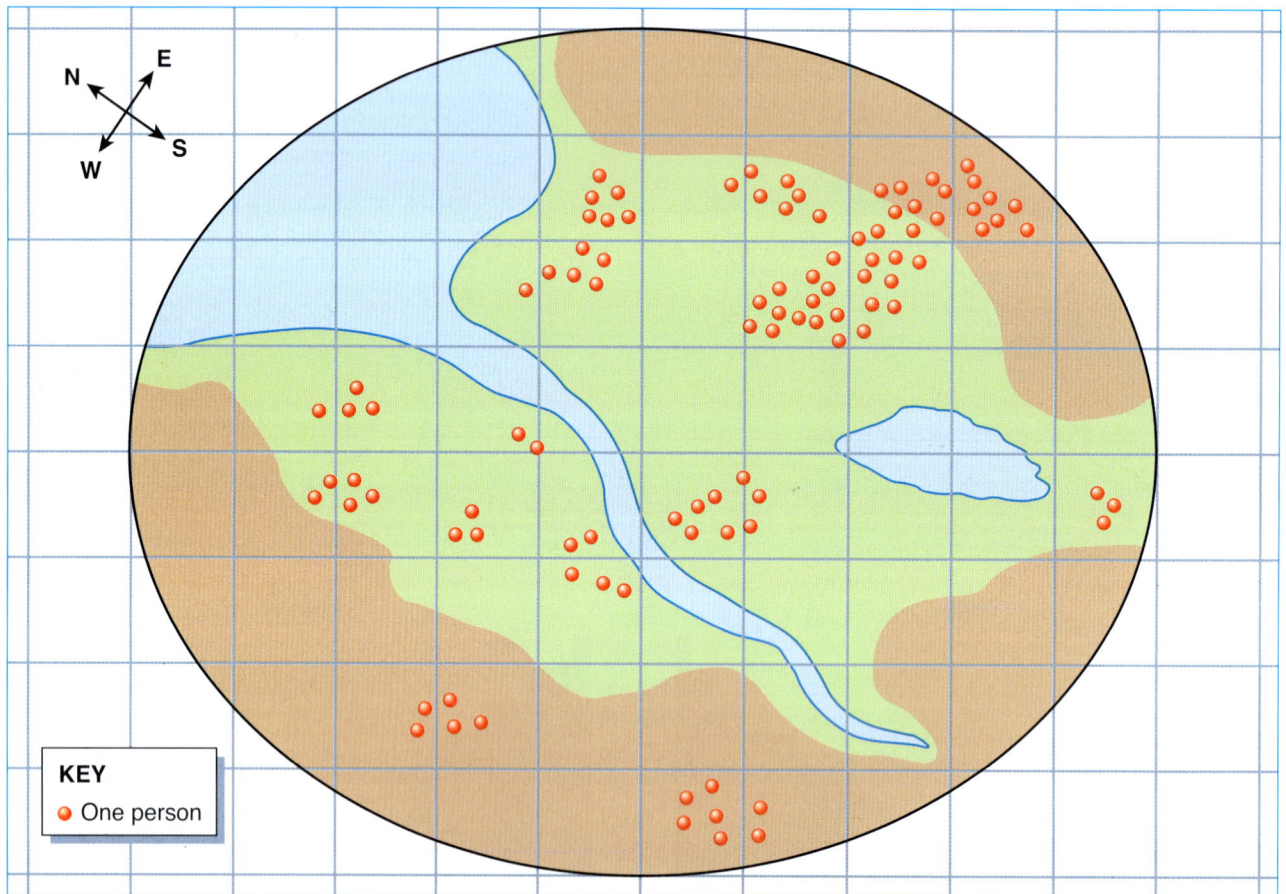

KEY
● One person

activity...

1 Describe the population distribution in Earth Village in map **E**. Copy and complete the paragraph below, using these words:

densely   east   concentrated-
unevenly   sparsely   north   fairly

People in Earth Village are _unevenly_ spread. Most people are _concentrated_ in the ___east___ of the village. This area is _densely_ populated. The area near the coast, in the ___north___ of the village is ___fairly___ densely populated. The rest of the village is _sparsely_ populated. Fewer people live in these areas.

... Ur welcome, 418 child

Map **E** is divided into squares. The number of people in each square is the population density. Some squares are crowded: they have a high population density. Some squares have few people: they have a low population density.

High population density

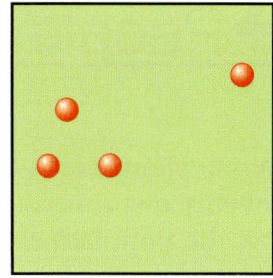

Low population density

2 You are going to make a map of population density for Earth Village.
   a) Draw a grid like this, showing the squares in map **E** (or your teacher will give you a copy).
   b) Count the number of people in each square on map **E**. The number in each square is the population density. Write the number in the squares on your grid in pencil.
   c) Give your map a key like the one here. Colour the key to show the different population densities. The more people, the darker you should make the colour.
   d) Colour your grid to show the population density in each square. This is your population density map.

KEY
People per square

☐ 0–1	☐ 4–5	☐ 8–9
☐ 2–3	☐ 6–7	☐ 10 or more

aim high...

3 Compare the population distribution in Earth Village on map **E** with this map of the physical geography of the area.
   a) Explain the population distribution that you described in activity **1**. Why are some areas densely populated? Why are some areas sparsely populated?
   b) Think of other factors, that could affect population distribution, that you cannot see on the physical map. For example, *wind direction*. Make a list.

KEY
- ▨ Land over 100m
- ～ River
- ◠ Lake
- ⸜ Marshland

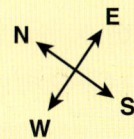

N E S W

→ People everywhere

Like people in Earth Village, people in the real world are unevenly distributed (map **F**). The colours on the map show the population density in each part of the world.

Andes Mountains, South America

Sahara Desert, Africa

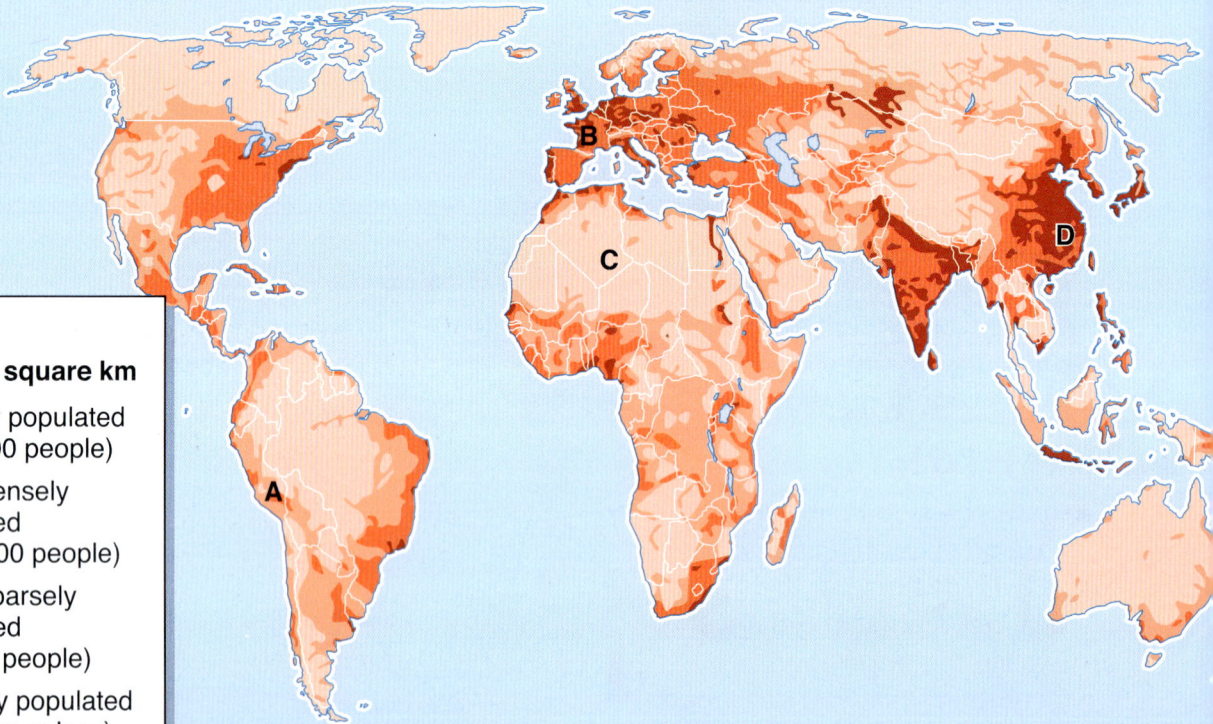

KEY

people per square km

- Densely populated (over 100 people)
- Fairly densely populated (10 to 100 people)
- Fairly sparsely populated (1 to 10 people)
- Sparsely populated (1 person or less)

F

World population distribution

France, Europe

China, Asia

12 in Europe

61 in Asia

5 in N. America

8 in S. America

13 in Africa

1 in Oceania

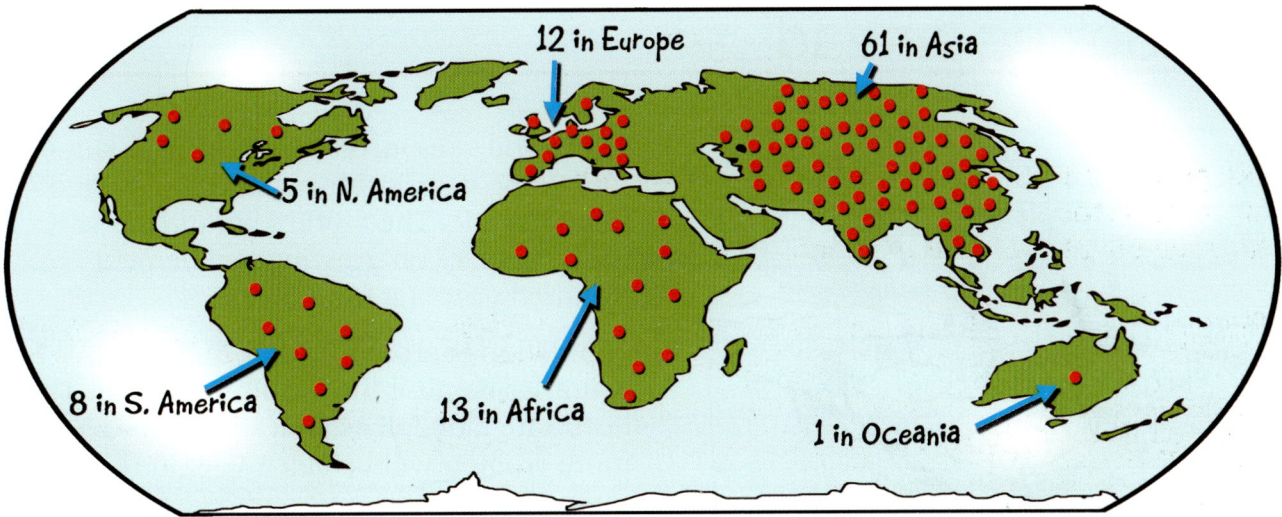

This is where people in Earth Village live.

activity...

1 Look at map **F** and the photos around it.
   a) Describe the population density you
   can see in each of the photos. Choose
   one of these phrases for each photo:

   densely populated   fairly densely populated
   fairly sparsely populated   sparsely populated

   b) Now, match each photo with one of
   the letters – **A, B, C, D** – on the map.
   Use the key to help you.
   c) Complete a table, like this:

Place	Letter on the map	Population density
Andes Mountains		
Sahara Desert		
France		
China		

2 Describe population distribution around
the world in map **F**. Write a few
sentences (similar to what you wrote in
activity 1 on page 12). Here are some
sentence starters you could use.

   The world's population is …
   Most people live …
   Few people live …

aim high...

3 What makes some areas densely
populated and other areas sparsely
populated?
The photos around map **F** should give
you lots of clues.
   a) Draw a large table with two columns,
   like this:

Densely populated	Sparsely populated
warm climate	cold climate

   b) List in your table all the things that
   would make an area densely
   populated, or sparsely populated.
   One is done for you. You should think
   about: temperature, rainfall, relief,
   location, vegetation, soil and farming,
   water supply, hazards and job
   opportunities.

15

→ Looking ahead

> Plenty of fish in the sea? I don't think so!

There is a problem in Earth Village. Resources are starting to run out. For example, fish.

The world's resources are starting to run out too. Some people think that there are too many people. The world's population is likely to carry on growing for the next hundred years (graph **G**).

Population is growing faster in poor countries because the birth rate is high. The birth rate may fall as education improves, as more women will be able to get jobs.

Population growth is much slower in rich countries. In a few rich countries the population has started to fall.

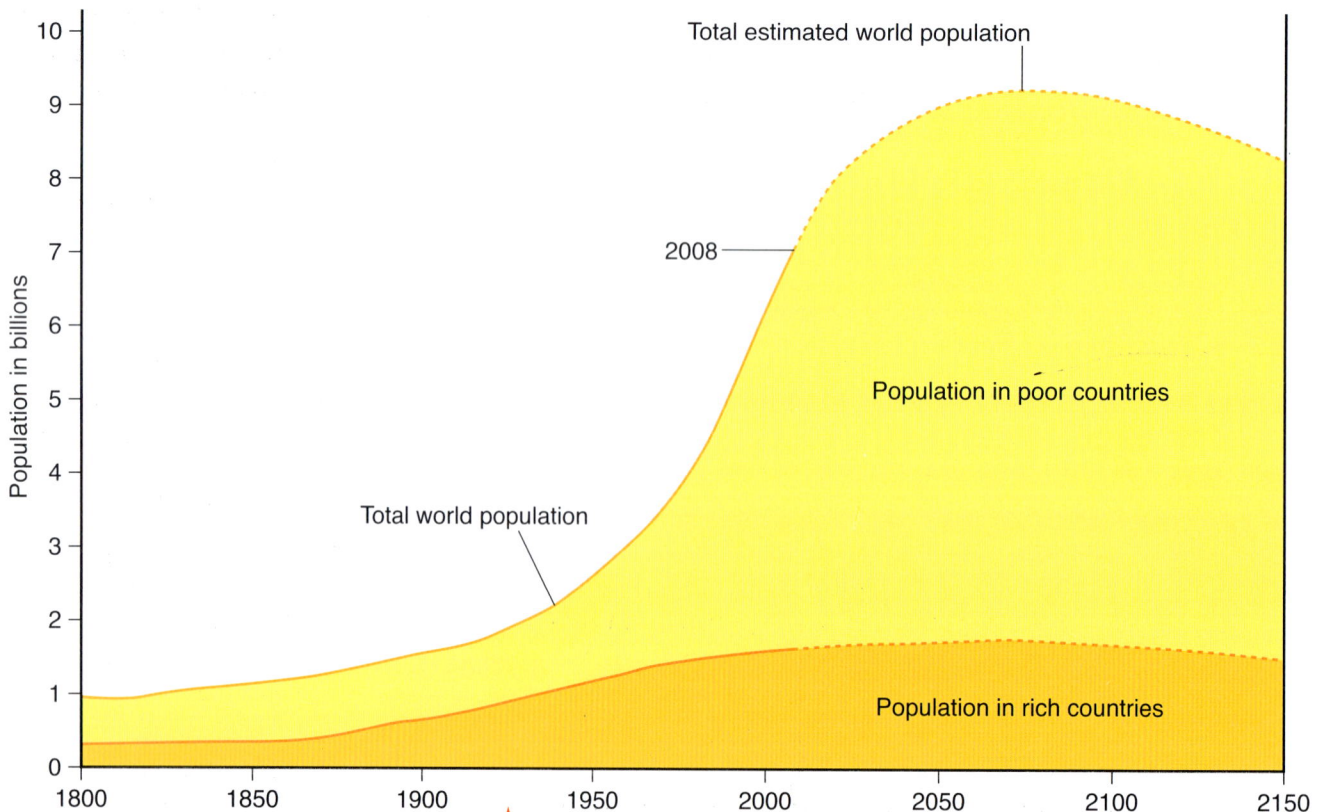

Total estimated world population

2008

Population in poor countries

Total world population

Population in rich countries

G How the world's population is likely to grow until 2150

→ Ecological footprint

Other people think the problem is not the number of people, but the amount of resources that each person uses.

The more resources we use the greater our impact on the Earth. You can measure your own impact on the Earth from your ECOLOGICAL FOOTPRINT.

Your normal footprint is the mark your shoe leaves when you walk on the ground. Your ecological footprint is the mark your lifestyle leaves on the world. The more resources you use, the greater your ecological footprint.

◄ This is the footprint for someone in the UK. If everyone in the world lived like us we would need three planets to support the population!

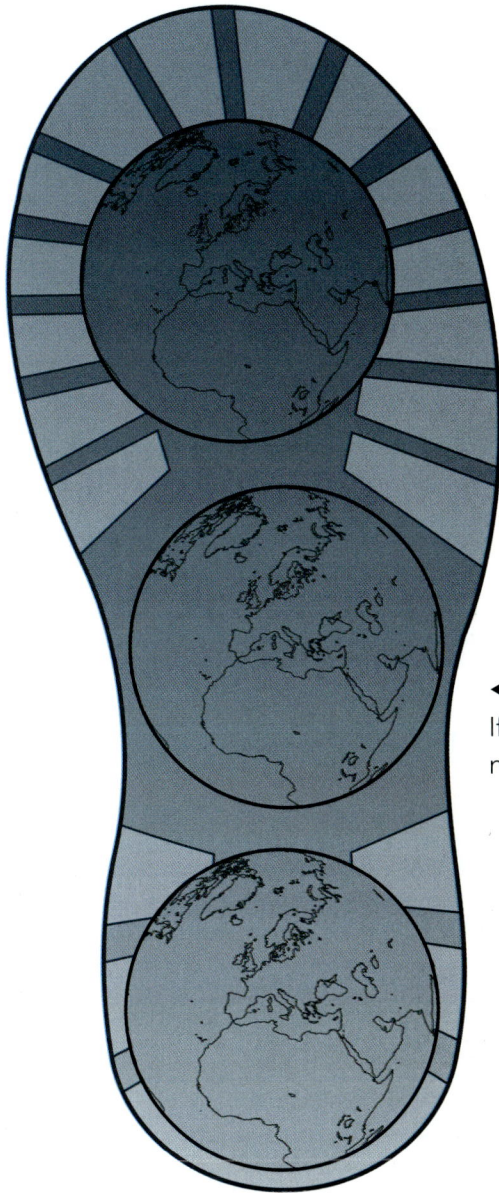

◄ This is the footprint for someone living in Kenya. If everyone in the world lived like this we would have more than enough resources on our planet.

activity...

1 Look at graph **G**.
  a) What was the world population in 2008? What could it reach by the year 2100?
  b) How will the number of people in,
     i) rich countries,
     ii) poor countries,
     change by 2100?
  c) What is likely to happen to world population after 2100?
2 Work out the size of your ecological footprint.
  a) Go the Earthday website at www.myfootprint.org Click on 'Skip intro'. Choose 'English' then click on 'And enter here'.
  b) Answer the questions to work out your footprint. If everyone lived like you, how many planets would we need?

discuss...

3 Work with a partner. Study all the information on these two pages. One of you should imagine that you live in Kenya. Discuss these questions:
  a) Do you think the world's biggest problem is the number of people or the amount of resources each person uses?
  b) How could the country you live in affect your opinion?

➡ Can the Earth cope with any more people?

When it comes to talking about population growth there are two types of people in the world: pessimists and optimists.

Pessimists believe that the world cannot cope with any more people. They think that the world is heading for disaster as the population grows, and we use more and more resources.

Reasons to be pessimistic

Over half the world's population are under 30 years old. Many of these people are yet to have children, so it will take a long time for the population to stop growing.

The world is getting warmer because we add carbon dioxide to the atmosphere by burning fuel. The more people there are, the worse this is going to be. The warmer climate causes problems like flooding and drought.

It could take another 150 years for the world's population to stop growing. By then there could be 10 billion people, or even more.

Nearly half of the world's population are farmers. As the population grows there is less land to go around.

Many poor countries are struggling to support their populations. Their populations are also the ones which are growing most rapidly.

Millions of people now face a severe shortage of water. Most of them live in poor countries where the population is still growing fast.

The world is likely to run out of oil in the 21st century. If the population grows, and we use more energy, it will happen even sooner.

■ your final task...

You are going to write a speech for a debate 'Can the world cope with any more people: yes or no?'

Study the information on these pages. Decide which side of the debate you will be on.

Look back over the whole unit. What other evidence can you find to support your view? (Don't forget the idea of Earth Village. It may help you to understand more about the world's population.)

Optimists believe that the world can cope with more people. They think that the population will stop growing and we'll find better ways to use the resources we've got.

So, what are you – an optimist or a pessimist?

Reasons to be optimistic

Population growth has always led to better farming methods, so that more food can be grown. It will be the same in the future

The population is falling in some countries where birth rates are lower than death rates. This is likely to happen in more countries in future with better health care and education.

The world already produces enough food for everybody. We just need to make sure that it reaches the people who need it.

Education is improving around the world. It is known that as more people are educated, so birth rates fall.

Around the world, women are getting better educated. They will be more likely to plan their families and to delay having children if they go out to work.

Technology can help us to use our resources more sensibly. For example, there are alternative sources of energy to oil and coal.

Write a speech for the debate, to last about one minute. You can use a writing frame to help you. Use evidence that you find to support your view. For example, if you are an optimist (using evidence from page 10): Birth rates in Kenya will fall by 2050 and the population will eventually stop growing.

Listen to the speeches that other pupils make. Think of questions to ask.

At the end of the debate your class could have a vote. Are you optimistic or pessimistic about the future?

Landscape detective

What can you learn about a place by investigating rocks?

▌▌ coming up...

Have you ever wondered why every landscape looks so different? The answer lies under your feet!

The rock beneath the ground makes each landscape unique. Even in cities, many buildings are made from rocks. If we dig a bit deeper, rocks also give us clues about how the landscape was formed.

You are a landscape detective. You will look for clues about how the landscape is formed. You will start on a trail around your local town, and then end up in the Yorkshire Dales.

■ your final task...

At the end of the unit, you will plan a trail around the Yorkshire Dales. This one will be a bit more adventurous!

IGNEOUS ROCK is made of crystals. You can see the crystal pattern in most igneous rocks. This is granite, a common igneous rock often used for kerbstones and shop fronts.

SEDIMENTARY ROCK is made from sediment – bits of clay, sand, stones or shells, all stuck together. Limestone is a common sedimentary rock used for building. This is travertine, a type of limestone, used on the front of McDonald's restaurants.

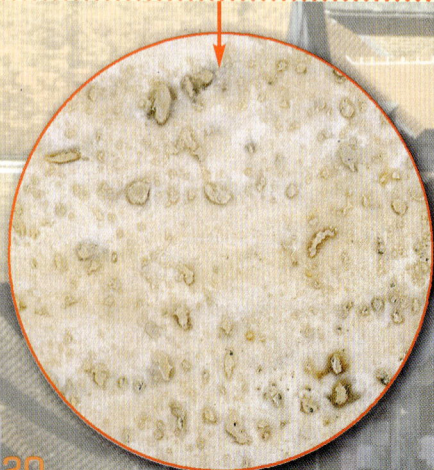

investigate...

1 Follow a trail around your local town centre. Your teacher may give you a trail to follow or you can make up your own trail on a map.
   a) How many different kinds of rock can you find? Draw a sketch (or take a photo) and write down anything you notice about each rock you find.
   b) Mark where you find each rock on a map.

You'll be surprised at how many rocks you can find in your street! Look out for four types of rock: igneous, sedimentary, metamorphic or made by people.

Rock made by people is used for many building jobs. For example, bricks are made by heating clay. Concrete, glass and ceramic tiles are man-made too.

METAMORPHIC ROCK is more difficult to recognise. It is formed from igneous or sedimentary rock that has been changed by heat and pressure. Slate is a metamorphic rock that was once clay. It is used for roof tiles.

aim high...

2 When you come back to the classroom you can try to identify the names of each rock. You could use rock identification tables on the 'About:Geology' website at: www.geology.about.com/library/bl/blrockident_tables.htm to help you. Click on any rock in the table to see a photo. Don't worry if you can't find the exact rock. Being a landscape detective isn't always easy!

21

➡ The rock cycle

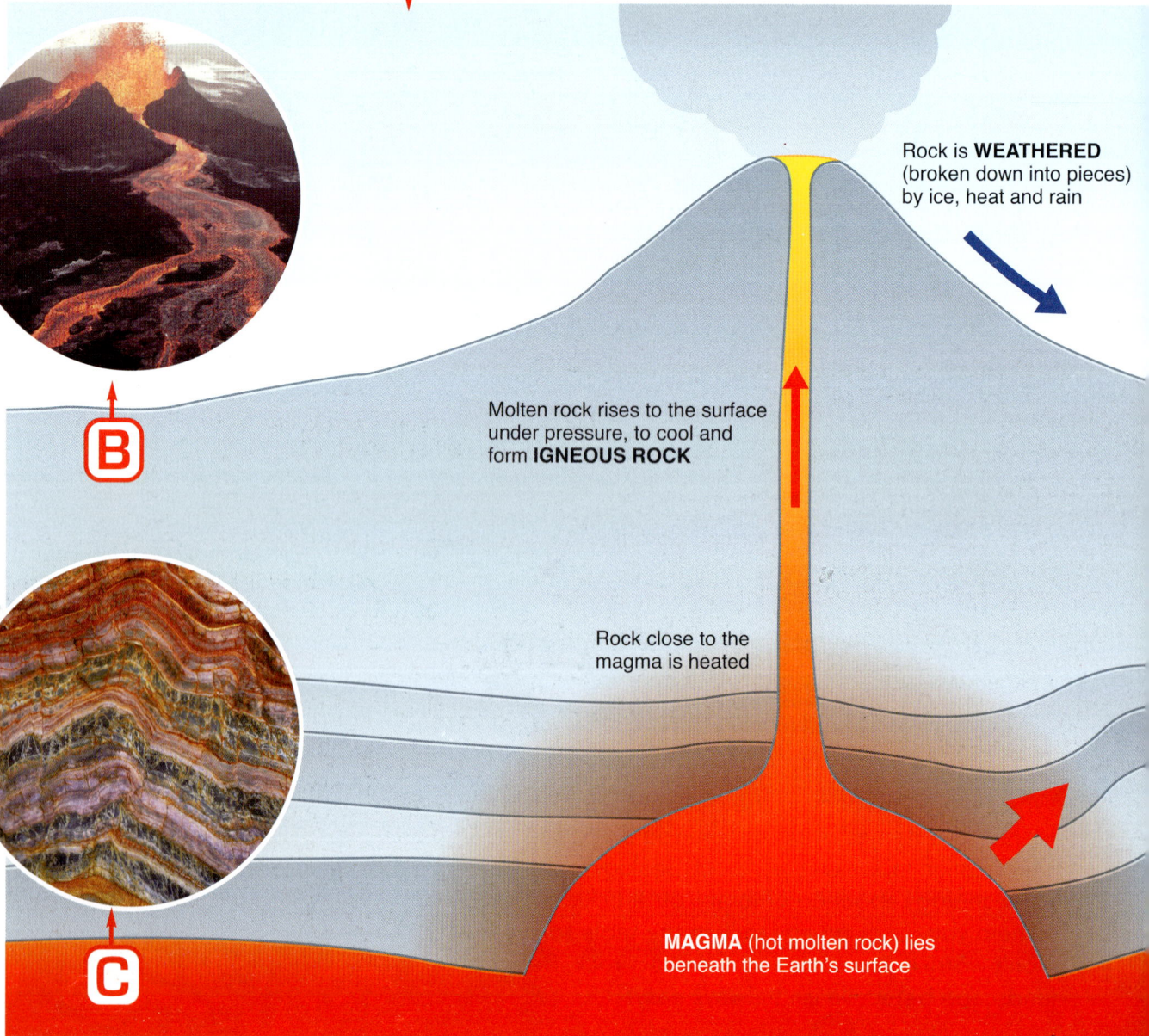

Most of what you study in geography is about what happens on the ground. A geologist is more interested in the rocks **below** the ground, called GEOLOGY.

Why do volcanoes occur? Where do fossils come from? How do rocks get folded? These are the type of questions geologists try to answer. One way to understand rocks is to look at the ROCK CYCLE.

A The rock cycle

B

C

Rock is **WEATHERED** (broken down into pieces) by ice, heat and rain

Molten rock rises to the surface under pressure, to cool and form **IGNEOUS ROCK**

Rock close to the magma is heated

MAGMA (hot molten rock) lies beneath the Earth's surface

activity...

1 Study drawing **A** carefully.

a) Make a large diagram of the rock cycle, like the one below, in your book.

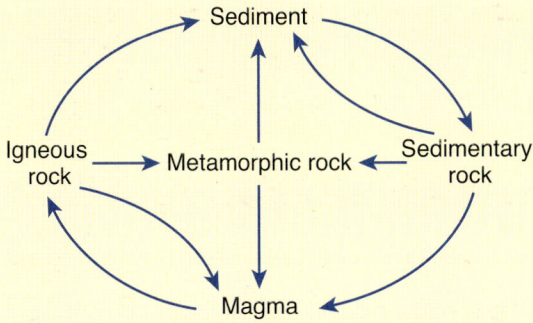

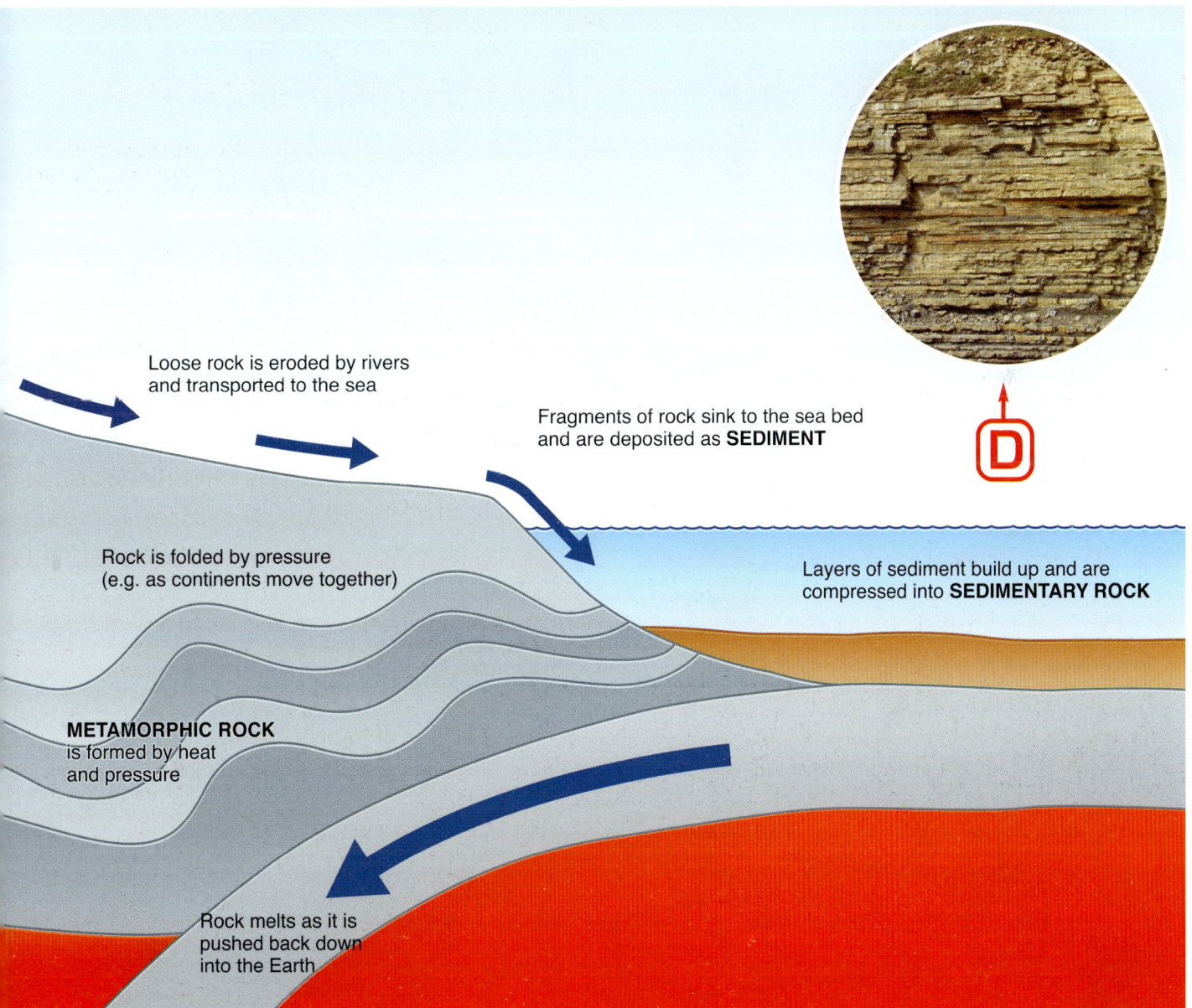

Sediment

Igneous rock → Metamorphic rock ← Sedimentary rock

Magma

b) Add the following labels to the arrows on your diagram. You can use the same label more than once.

★ melting ★ cooling
★ compression ★ heat + pressure
★ weathering, erosion + deposition

2 Look at the photos.

a) What kind of rocks are shown in each of the photos **B**, **C** and **D**: igneous, sedimentary and metamorphic?

b) Explain how each rock is formed.

Fragments of rock sink to the sea bed and are deposited as **SEDIMENT**

Loose rock is eroded by rivers and transported to the sea

D

Rock is folded by pressure (e.g. as continents move together)

Layers of sediment build up and are compressed into **SEDIMENTARY ROCK**

METAMORPHIC ROCK is formed by heat and pressure

Rock melts as it is pushed back down into the Earth

➡ Layer upon layer

If you look at an ordinary landscape, with a bit of imagination, you can see the rocks underneath. Most of the landscape in England is made of layers of sedimentary rock. Sedimentary rock can be rather like slices of bread and butter. The slices are piled up on top of each other in layers, so that the oldest slices are at the bottom and the newest slices are at the top. Sometimes the slices begin to tilt over.

E Chalk downs in England, showing the rock below the surface

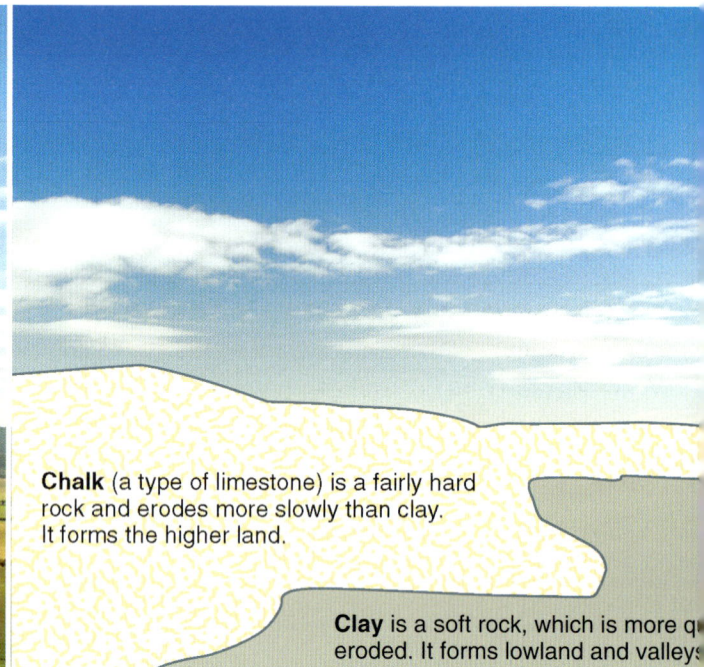

Chalk (a type of limestone) is a fairly hard rock and erodes more slowly than clay. It forms the higher land.

Clay is a soft rock, which is more q[u]ickly eroded. It forms lowland and valley[s]

activity...

1 Look at photo **E**.

a) Complete a large copy of this sketch to show the landscape in the photo. Label the 'chalk' and the 'clay'.

b) Now copy this simple geology map showing the same area. Label the two rocks.

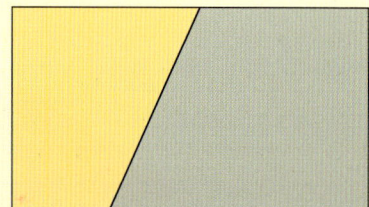

KEY

Rock	Date
	(millions of years before the present)

Sedimentary

Alluvium	2
Sands and clays	
London Clay	70
Chalk	
Greensand	
Weald Clay	
Oxford Clay	
Jurassic Limestone	
Liassic Beds	
Marl and Sandstone	220
Permian Marl	
Magnesian Limestone	
Coal Measures	
Millstone Grit	
Carboniferous Limestone	
Old Red Sandstone	
Shales and slates	
Slates and volcanic rock	600
Metamorphic	
Igneous	

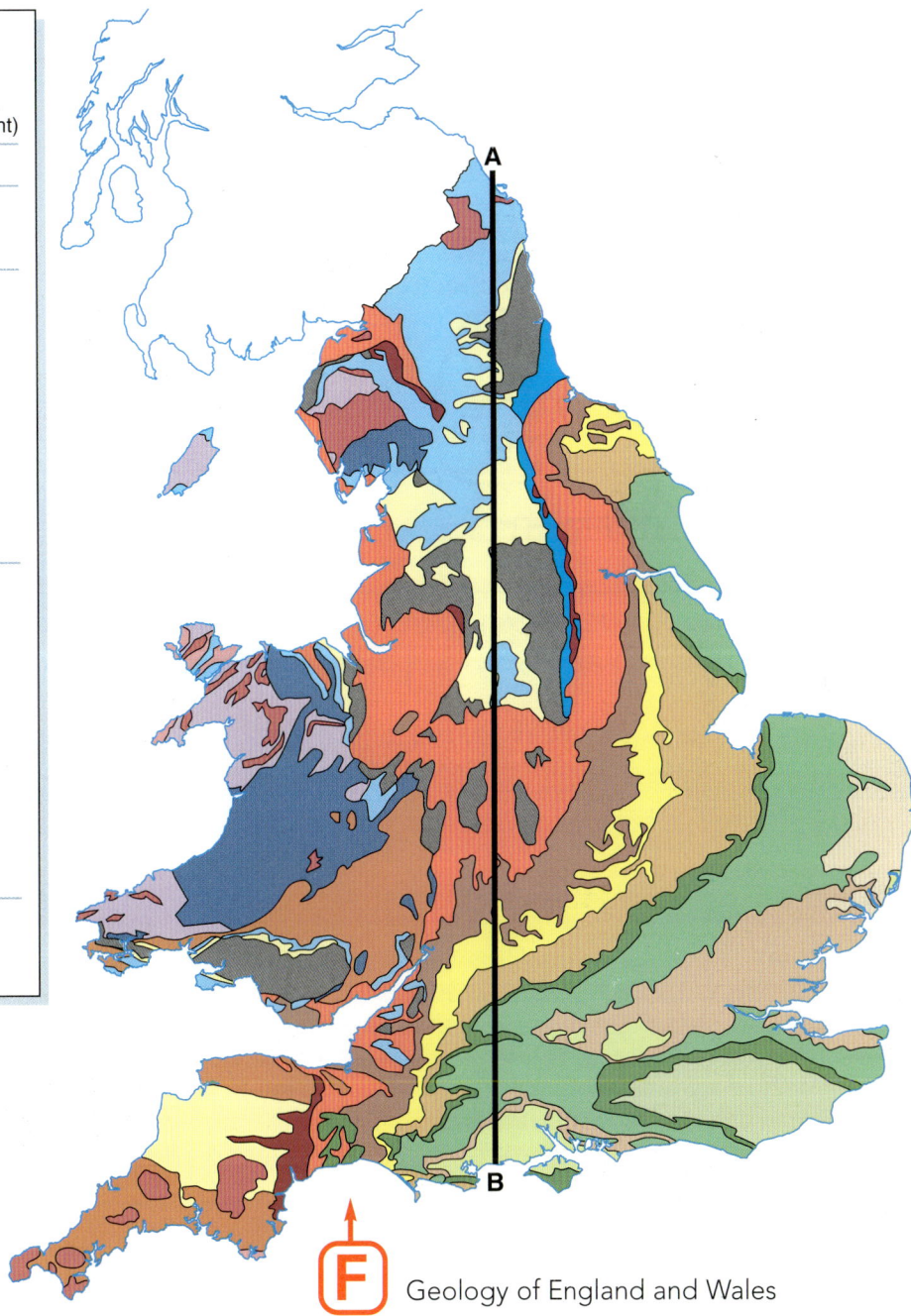

0   200 km

F Geology of England and Wales

2 Look at map **F**. It shows the geology of England and Wales.
  a) What rocks would you cross if you travelled from **A** to **B**? List them in order.
  b) Which is the oldest rock in your list? Which is the youngest?

3 Compare map **F** with a relief map in your atlas showing the high areas of England.
  a) Name three rocks that you find in hilly areas. What does this tell you about the rocks?

  b) Name three other rocks that you find in lowland areas. What does this tell you about the rocks?

aim high...

4 Write a paragraph to explain the connection between landscape and geology in England and Wales.

→ From the seabed to mountain top

When climbing in the hills of the Yorkshire Dales in the north of England I found a FOSSIL. It was sticking out of the limestone. A fossil is the remains of a dead plant or animal preserved in rock. This one was a sea creature that lived millions of years ago! You could find similar creatures living today in coral reefs in shallow tropical seas. Why is a tropical sea creature on the top of a hill in the north of England?

activity...

1 You are going to solve this mystery:
'How does a tropical sea creature come to be on top of a hill in the north of England?'
Read the sentences below – they are muddled up. Match each sentence with one of the drawings in the storyboard. (Your teacher may give you a copy of the sentences to cut out.) Write (or stick) the sentences in the correct order into your book.

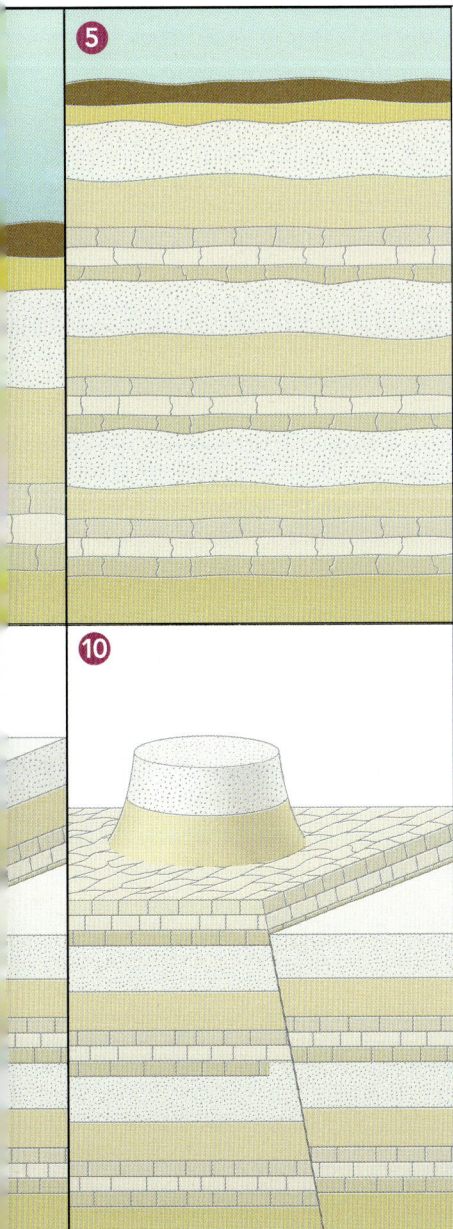

a) 280 million years ago, Africa collided with Europe. Rock was pushed up to form mountains.

b) Small sea creatures built their skeletons and shells from calcium carbonate. When they died they sank to the seabed.

c) Under extreme pressure the rocks cracked along a fault. This exposed a layer of buried limestone at the surface.

d) Layers of mud and sand covered the limestone, turning into shale and sandstone.

e) 350 million years ago Britain lay close to the Equator. Shallow tropical seas covered much of northern England.

f) All the while the continents were moving. It took millions of years for Europe to move from the Equator to its position today.

g) Layers of shells and chalky mud built up. They were squeezed and hardened until they eventually turned into limestone.

h) Sea level rose and fell many times over millions of years. The cycle was repeated over and over – limestone, sandstone, shale ... limestone, sandstone, shale.

i) The layers of rock were folded upwards under pressure to form a dome.

j) The layers of shale and sandstone were eroded to expose more limestone at the surface.

aim high...

2 How are a tropical sea and a limestone hill in the north of England connected?
Write one or two sentences in your own words to explain the connection between the two places.

→ Limestone's hidden depths

On the map that I use to explore the Yorkshire Dales streams seem to disappear in one place, then re-appear somewhere else. Why? With a rock like limestone, what is going on below the ground is very important ...

aim high...

3 A lot of bottled water comes from springs in limestone areas. Can you suggest why? Produce an advert for water from the Yorkshire Dales, telling people why it is good to drink.

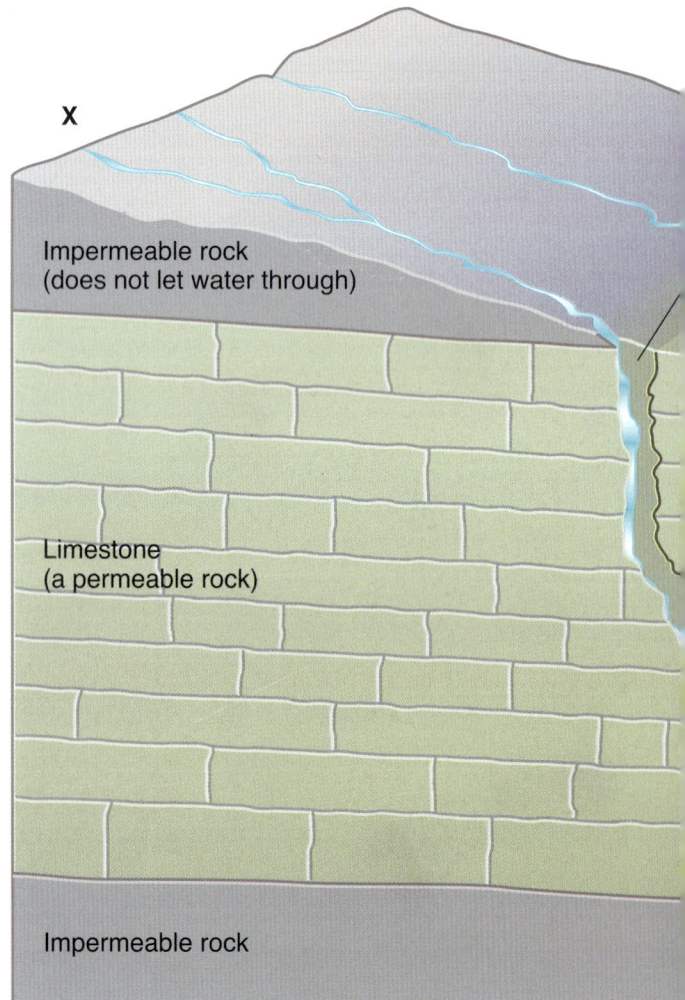

activity...

1 Study photo **H** and the information in **I**. How was the cave formed? Write one or two sentences to explain.
2 Study drawing **J**.
Imagine a water droplet flowing from **X** to **Y** on the drawing. Write a story to describe its journey through the limestone. Mention all the bold words on the drawing.

X

Impermeable rock
(does not let water through)

Limestone
(a permeable rock)

Impermeable rock

H Gaping Gill, a limestone cavern in the Yorkshire Dales.

s of seashells and chalky mud build the seabed. They are made from IUM CARBONATE, a chemical that ves slowly in water.

The layers (called **STRATA**) harden into limestone. The rock shrinks forming cracks, called **JOINTS**. The lines between the strata are **BEDDING PLANES**.

Water seeps into the joints and bedding planes. As the limestone dissolves these get wider and grow into **POTHOLES** and **CAVES**.

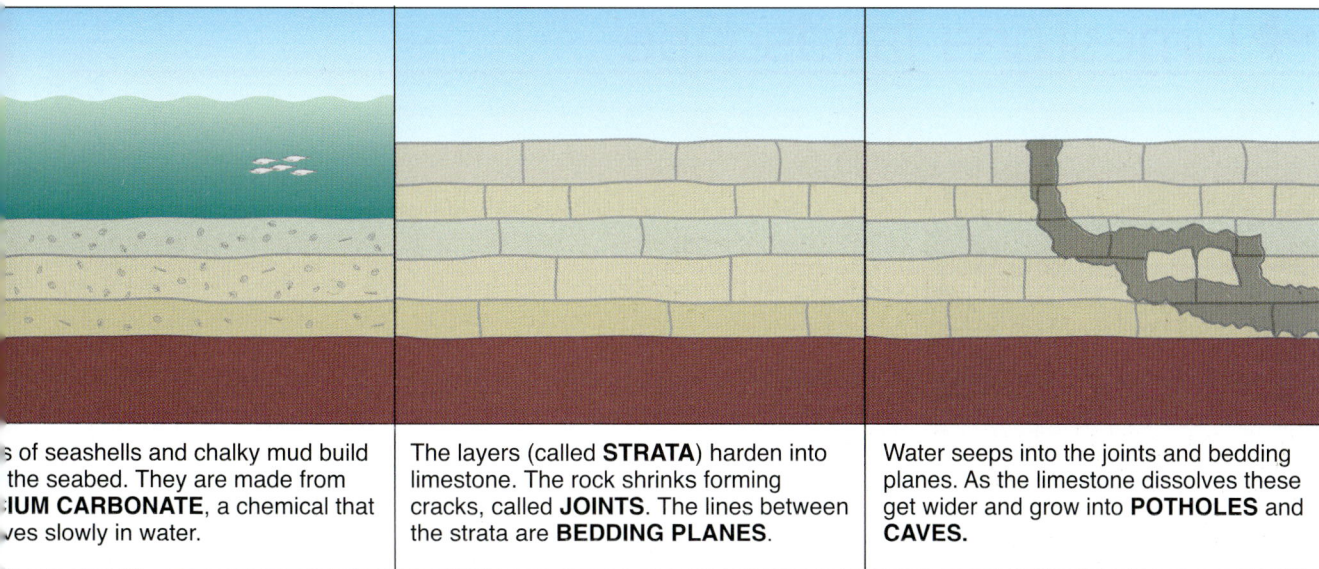

I How caves are formed in limestone

J Cross-section through limestone landscape

SWALLOW HOLE where the water disappears into the rock. Holes without a stream are **POT HOLES**.

DRY VALLEYS were formed in the Ice Age when the ground froze and streams flowed over the surface. Today the streams flow underground.

LIMESTONE PAVEMENT is formed where water dissolves limestone on the surface to widen the joints.

ater dripping from the cave deposits lcium carbonite (like the fur in a kettle). ver many years it forms **STALACTITES** ing to the ceiling) and **STALAGMITES** row from the ground). Over thousands years more limestone dissolves and ves grow into **CAVERNS**.

Limestone is a hard rock that erodes slowly. It forms steep cliffs called **SCARS**.

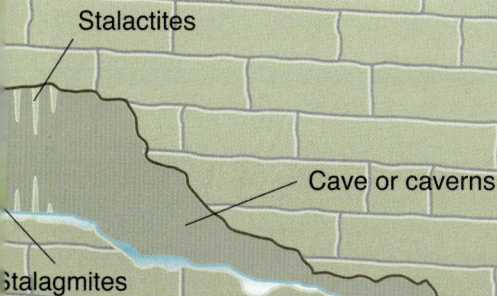

Stalactites

Cave or caverns

A **SPRING** where a stream emerges from the limestone to flow over the surface again.

talagmites

Y

→ Limestone landscape

In the Yorkshire Dales all around you is evidence of the rock that lies under your feet. You don't need to be the smartest detective to work out what this rock is. There are limestone buildings, limestone walls – even limestone pavement!

The Yorkshire Dales is a national park. Lots of people come here to enjoy outdoor activities, but there are other people who live and work here.

activity...

1 Look at all the photos on this spread. They were all taken in the Yorkshire Dales.

a) Identify each of these features of limestone landscape in the photos. You can use drawing **l** on pages 28 to 29 to help you.

> dry valley   village   pot hole   quarry
>
> spring   scar   limestone pavement

b) Which of these features are natural and which are made by people?

2 Choose four outdoor activities you could do in the Yorkshire Dales:

> fishing   hill-walking   football   hang-gliding
>
> canoeing   caving   skiing   climbing

a) Match the photos to each of your chosen activities. For example, *'You could hang-glide over a scar.'*

b) Explain why each photo is a good place to do this activity. For example, *'A scar is a good place to hang-glide because you can take off from the edge.'*

c) Explain why you did not choose the other four activities.

aim high...

3 From the evidence in these photos, would you describe limestone as:

a) a hard or soft rock?

b) a permeable or impermeable rock?

c) an igneous or sedimentary rock?

In each case only give evidence from the photos, not from what you already know.

→ Plan a limestone trail

A landscape detective can find out things about the landscape without even going there by using a map. This map tells you lots about the landscape. It is a map of Ingleborough in the Yorkshire Dales.

activity...

Look carefully at map **Q**. Match each feature numbered on the map with one of the photos on pages 30 to 31. Complete a table like this. One row is done for you.

Number on map	Grid reference	Photo	Limestone feature
1	758 730	M	Pot hole

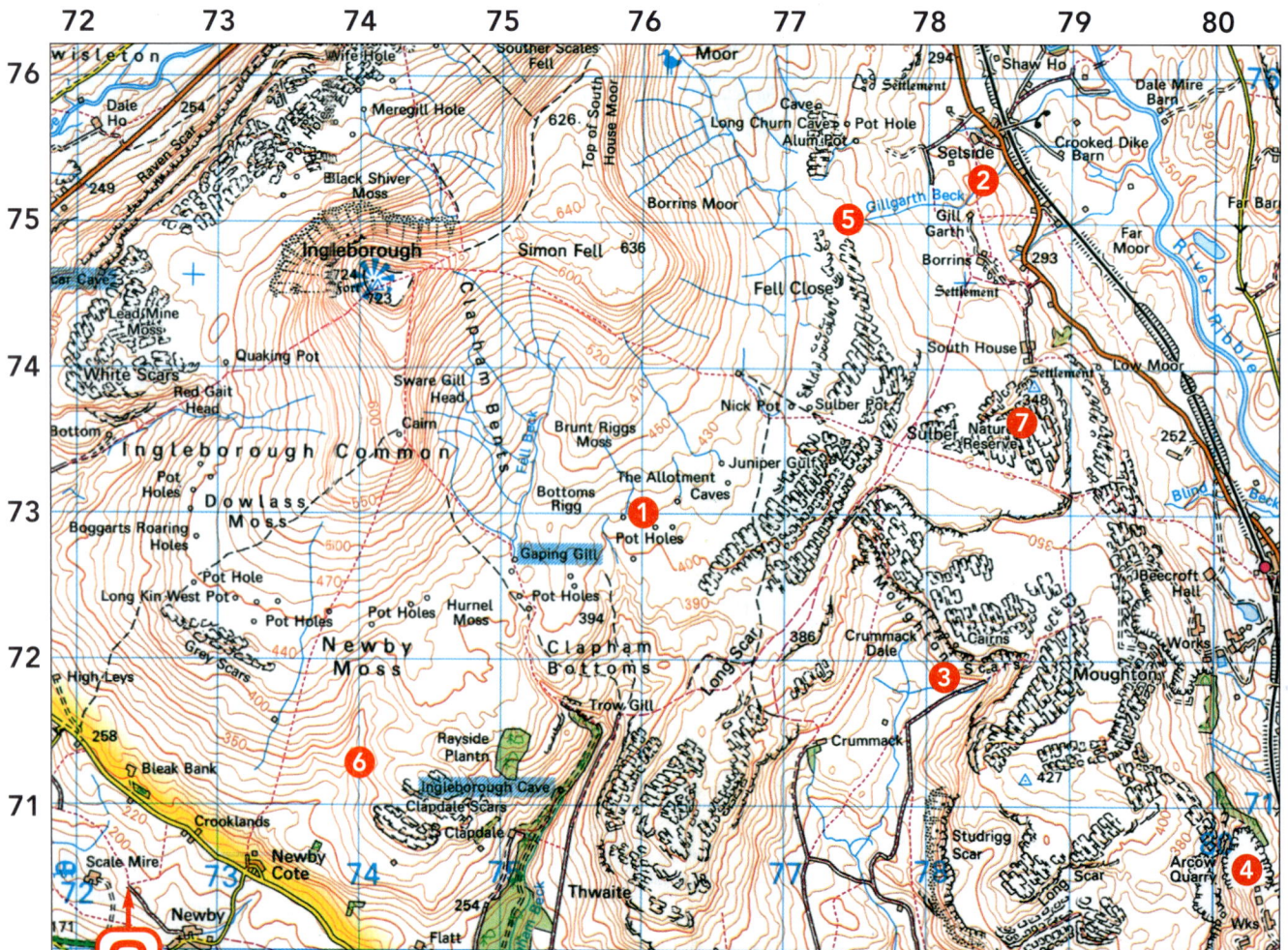

Q 1:50,000 OS map of Ingleborough

your final task...

Your task is to plan a limestone trail around Ingleborough using map **Q**. You must visit as many limestone features on your trail as you can. (It is like the trail you did for your town centre – but more adventurous.) However, you need to make sure that you bring all the equipment you need.

1 Think about the route you will take on the map. To get from one point to the next on your trail you may walk over hills, climb scars, dive down caves or even hang-glide. (The footpaths may help if you're walking, but otherwise you don't need to stick to them.)

2 Make a large copy of the table below in your books and fill it in:
   * Choose at least six points as stages in your route. Give grid references for the points at the start and finish of each stage.
   * Measure the distance and find the direction between the points. Do this for each stage of your trail.
   * Explain how you will get from one point to the next, listing the equipment you need. (The pictures here will give you some ideas.)
   * Describe the landscape you expect to see at each stage of the trail. The map has lots of information you could use (see also the photos on pages 30 to 31).

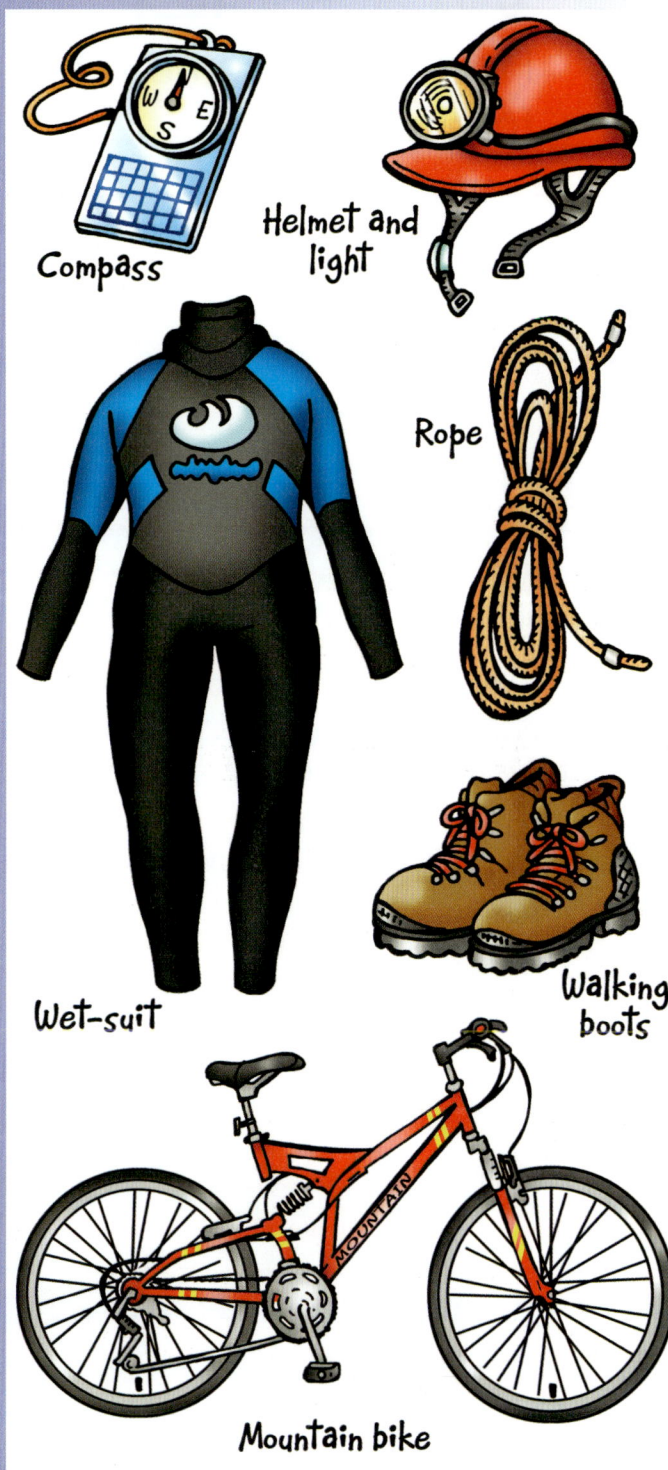

Compass

Helmet and light

Rope

Wet-suit

Walking boots

Mountain bike

From (grid ref)	To (grid ref)	Distance and direction	How I will travel and equipment	Landscape I expect to see

3 Here is the weather forecast.

How do I become a brilliant weather forecaster?

▌▌ coming up…

Have you ever watched the weather forecast on TV and thought, I could do that job? Unfortunately, it's not as easy as it looks. There's more to being a weather forecaster than you might think.

▌▌ through the unit…

In this unit you will learn some of the basic skills you need to do the job.

▌ your final task…

At the end, you'll get a chance to be a weather forecaster.

Let's go behind the TV screen and find out what goes into making the weather forecast …

Before you can forecast tomorrow's weather, you need to know what is happening now. All around the world, people observe and measure the weather.

All the information is sent instantly around the globe by modern telecommunications.

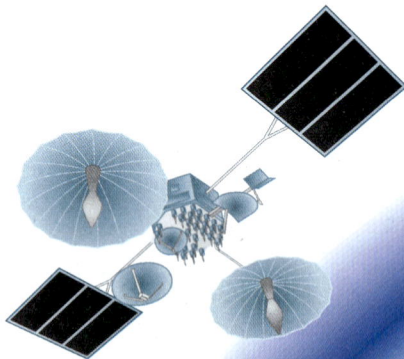

Balloons are sent up to find out what is happening high in the atmosphere.

Satellites send back images of the Earth from space. They show cloud patterns and changes in the atmosphere.

Ships report the weather at sea.

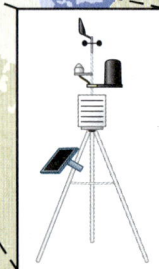

Weather stations on land record the weather.

activity...

1 Look at the weather forecast map of the British Isles.

a) Match these symbols with the weather they represent;

temp	rain
sun	cloud
sunny intervals	wind

Draw a key for the map.

b) Use the map to give the weather forecast for your region. Write one or two sentences.

aim high...

2 You are the weather forecaster. Write a script for the weather forecast from the map. Your forecast should include all parts of the British Isles. Use the symbols on the map to help you.

The information is brought together and turned into charts to show the weather patterns. (You will use a weather chart on page 46.)

Powerful computers do millions of calculations to forecast how the patterns will change over the next few days. But that still leaves one important job …

The weather forecaster has to make sense of all the charts and computer forecasts, and explain it to the rest of us – simply!

→ Weather check

Step 1 — *Record the weather*

Weather is what happens in the atmosphere from day to day. You can tell a lot about the weather just by looking (photo **A**). But to be able to forecast the weather, you also have to measure and record what happens. The instruments you need to measure the weather are in **B**.

How much rain has fallen?

How windy is it?

Where is the wind blowing from?

How sunny or cloudy is it?

What is the temperature?

What is the air pressure?

A A wet, windy day

activity...

1 Look at photo **A**.
  a) Describe the weather you can see just by looking at the photo.
  b) How accurate do you think your description is? Can you answer any of the questions on the photo? What questions can't you answer?
  (Don't worry if you're not sure what air pressure is! You'll find out on pages 42 to 43.)

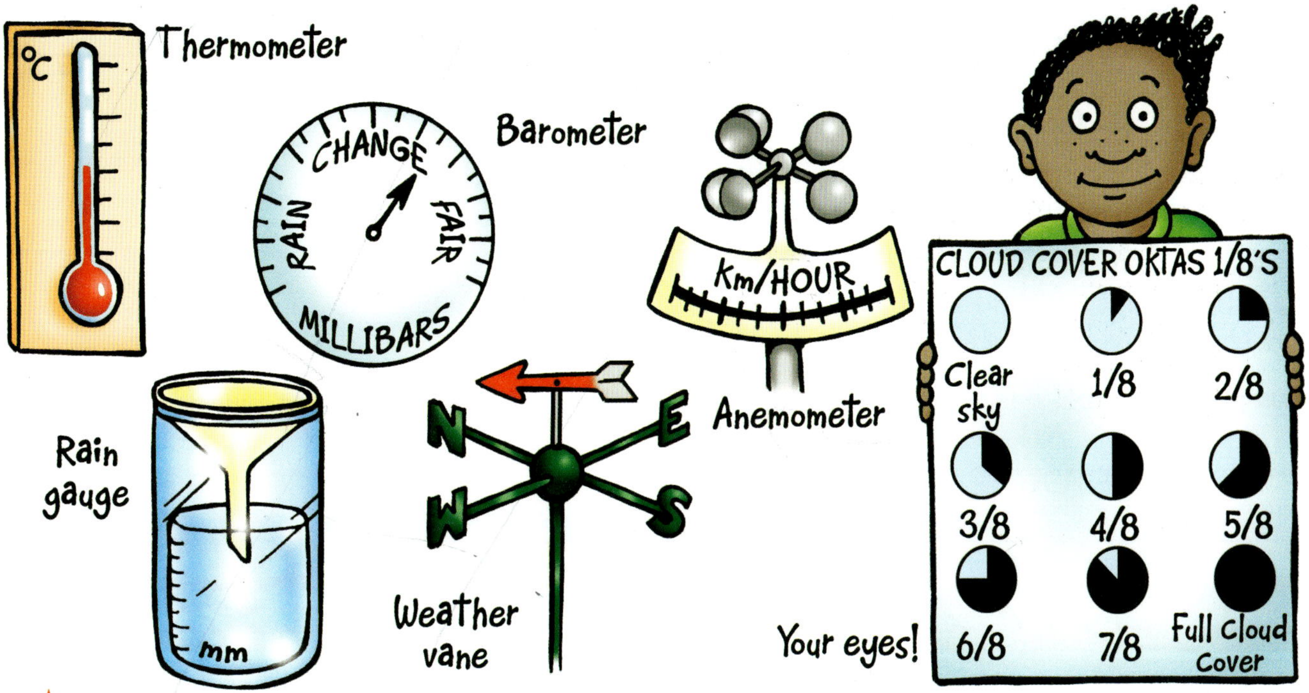

B Instruments used to measure weather

2 Look at the weather instruments in **B**. Then copy and complete the table on the right.

   a) Match each instrument with what it measures in column 2.

   b) Write the correct units to measure the weather in column 3. Choose from:

     km/hour   millibars   centigrade

     oktas   millimetres   compass point

Weather	Measured using …	Measured in …
Temperature		
Rainfall		
Wind speed		
Wind direction		
Air pressure	barometer	millibars
Cloud cover		

3 Measure and record the weather today. You will need to use instruments like those in box **B** at your school weather station (or, your school may have an automatic weather station).
If you don't have a weather station, or any instruments, you can get up-to-date measurements from the Met Office website at www.metoffice.gov.uk/education. Click on 'Weather data' and then 'Latest UK observations'. You can find measurements from the weather station that is nearest to you.

aim high...

4 Find the latest weather around the UK from webcams (cameras that give live pictures, twenty-four hours a day). Use the website, www.greatweather.co.uk. Go to 'Webcams'. Choose some places around the UK to investigate. Find the places in an atlas.

37

Clues in the clouds

Step 2   *Look at the sky*

You may have been in a cloud if you have climbed a mountain or been in a plane. How did it feel?

Clouds are made from tiny water droplets. But not all clouds are the same. It's almost as if they have their own characters! Each one gives us clues about the weather.

CIRRUS is cool (Ice cool in fact! It's freezing at 10,000 metres). He thinks he's so superior up there. Beware, rain is on the way.

CUMULONIMBUS is really bad tempered. When she's around the sparks will fly! Thunderstorms and heavy rain are sure to happen – so, take cover.

CUMULUS is bright and bubbly. She's everybody's friend – but, don't rely on her. Usually a sign of good weather, though you might get an odd shower.

STRATUS is soooo depressing. He blocks the sun and spreads gloom everywhere. And, if that doesn't dampen your spirits, the rain and drizzle will!

C

activity...

1 Look at the clouds in **C**.
Write a sentence to describe the appearance of each type of cloud. A bit of Latin could help you:

> ★ cirrus = hair   ★ cumulus = pile
> ★ stratum = layer   ★ nimbus = cloud

2 Look at the three types of rainfall in **D**. For each type of rainfall, note:
- which types of cloud are associated with it, e.g. cumulus
- what sort of rain you would expect, e.g. an odd shower

3 Look carefully at the drawings in **D**.
- **a)** What do all three types of rainfall have in common? Mention two things.
- **b)** What is unique to each type of rainfall? Mention one thing for each type.

aim high...

4 When you go through the clouds in a plane, you often feel it bumping up and down.
- **a)** What do you think causes this? (The clue is on these pages.)
- **b)** In which type of cloud will the bumping be worse? Why?

Clouds form when air rises and cools. Water vapour in the air condenses into water droplets. When the droplets get too heavy, they fall as rain. Air rises in three ways. Each of these ways produces a different type of rainfall.

D Three types of rainfall

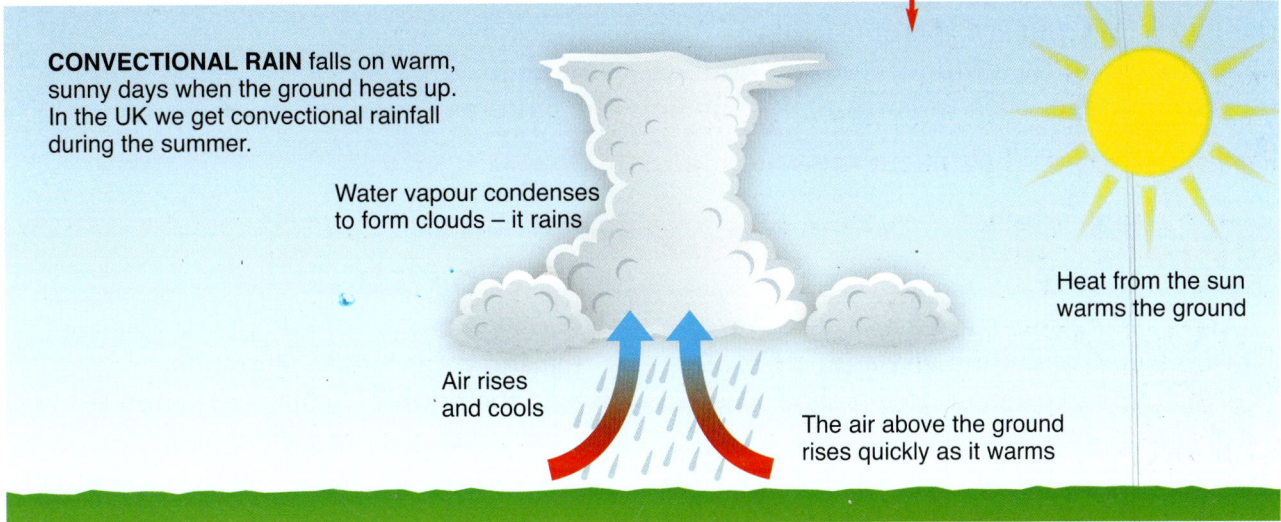

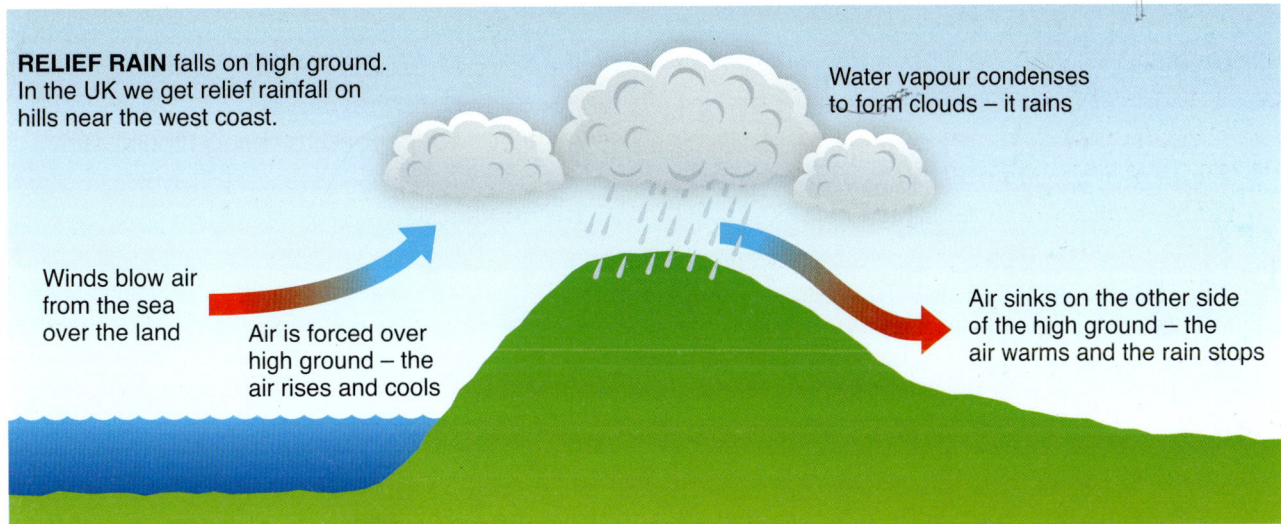

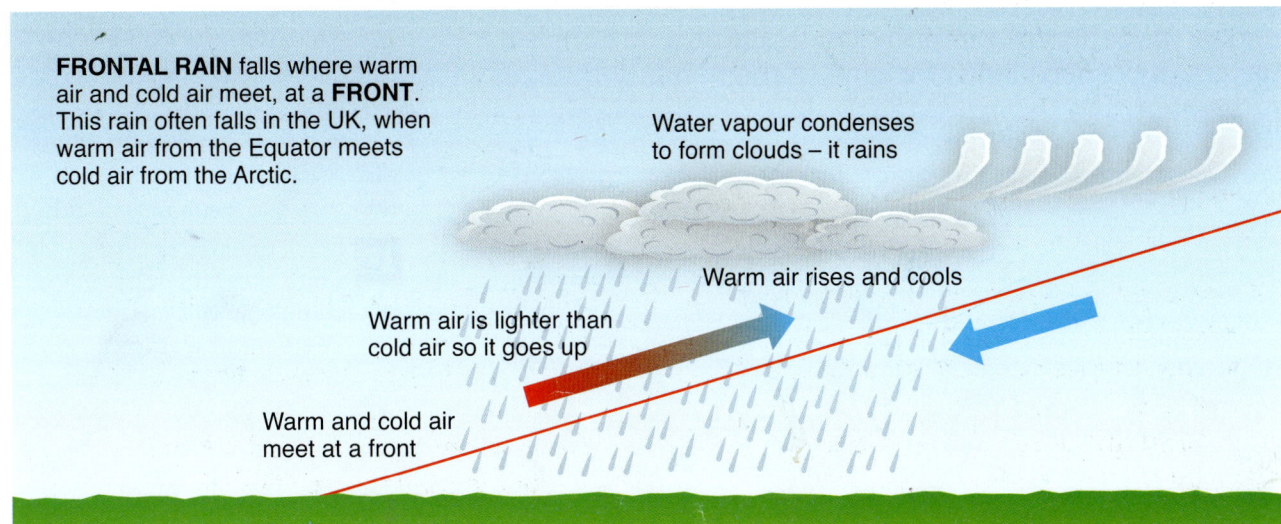

CONVECTIONAL RAIN falls on warm, sunny days when the ground heats up. In the UK we get convectional rainfall during the summer.

Water vapour condenses to form clouds – it rains

Heat from the sun warms the ground

Air rises and cools

The air above the ground rises quickly as it warms

RELIEF RAIN falls on high ground. In the UK we get relief rainfall on hills near the west coast.

Water vapour condenses to form clouds – it rains

Winds blow air from the sea over the land

Air is forced over high ground – the air rises and cools

Air sinks on the other side of the high ground – the air warms and the rain stops

FRONTAL RAIN falls where warm air and cold air meet, at a **FRONT**. This rain often falls in the UK, when warm air from the Equator meets cold air from the Arctic.

Water vapour condenses to form clouds – it rains

Warm air rises and cools

Warm air is lighter than cold air so it goes up

Warm and cold air meet at a front

➡ The view from space

Satellites make it easier to forecast the weather. High above the Earth they can see weather patterns that are impossible for us to see on the ground. Hour by hour, day and night, they send images back to Earth that show how the patterns are changing.

activity...

1 Look at satellite image **E** and compare it with a map of Europe on page 135.
What type of weather does it show in:
a) south of Spain **b)** north of Scotland?
In each case, explain how you can tell from the satellite image.

2 **a)** What types of cloud do you find at letters **X**, **Y** and **Z** on the satellite image?
b) What weather would you expect to find on the ground? (Look back at page 39.)

3 Explain why you find clouds at **Z**. Mention three things in your answer:
• the location of **Z**
• the time of year (August)
• the time of day (midday).

aim high...

4 In different parts of Europe, people are making plans for the afternoon.
• In the north of Scotland, Hamish is going to sea in a fishing boat.
• In northern Italy, Giovanni is climbing in the Alps.
• In southern Spain, Flavia is running a marathon.
If each person checked the satellite image before they started:
a) Who would need to change their plans, or not?
b) Explain why.

An infrared image is different from a normal photo. It shows heat rather than light. The coldest things show up white (like the cloud tops). The warmest things are much darker (like the surface of the land). Because infrared images don't need light, we can see them at any time, day or night.

Key	
■	Large dark areas show land or sea. There are no clouds so, by day, it is sunny.
	Large areas of white, or light grey, show stratus cloud. There may be rain.
	Areas of patchy white or grey show cumulus clouds.
	Small patches of bright white show the tops of cumulonimbus clouds. There may be storms.

E An infrared satellite image of north-west Europe, taken at midday in August. The coastline and lines of latitude and longitude are added to the image to make it easier to find places.

→ High pressure …

AIR PRESSURE is simply the weight of the air above pressing down on the Earth. Like someone on a trampoline – when the air is sinking it presses harder than when it is rising. Air pressure controls the weather:

- High pressure is when the air is sinking down on the ground – it often brings fine settled weather (photo **F**).

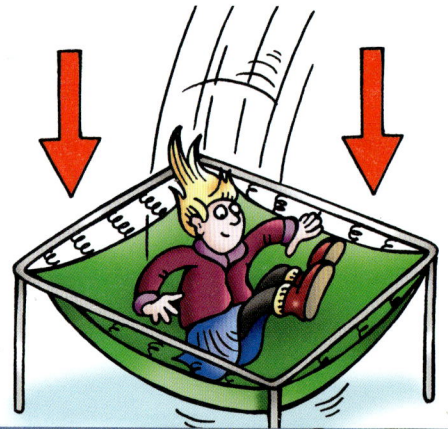

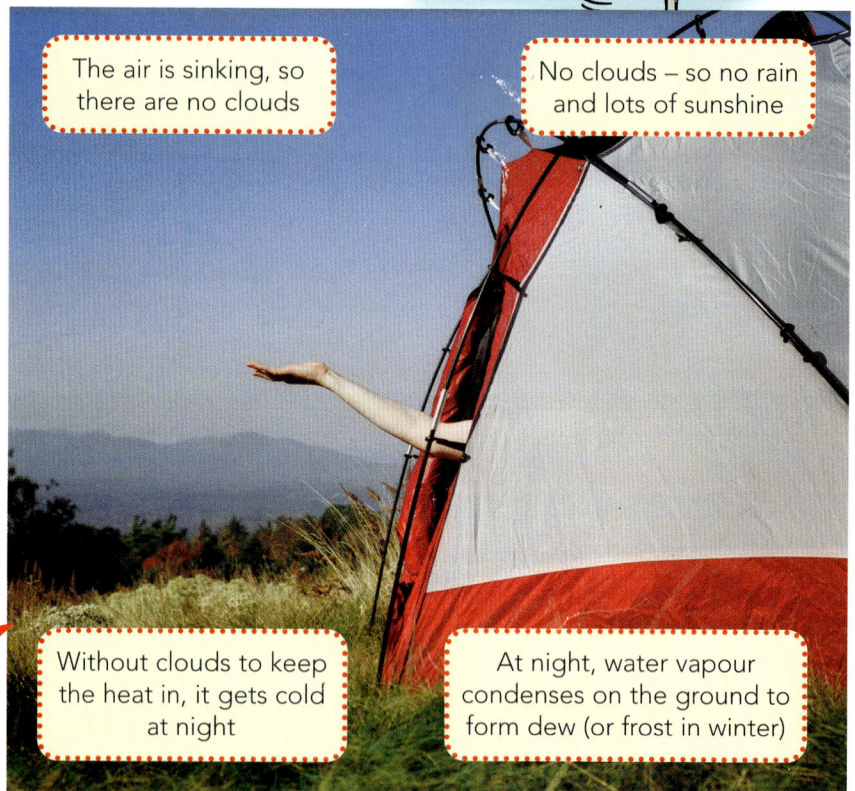

> The air is sinking, so there are no clouds

> No clouds – so no rain and lots of sunshine

> Without clouds to keep the heat in, it gets cold at night

> At night, water vapour condenses on the ground to form dew (or frost in winter)

F   Camping in high pressure

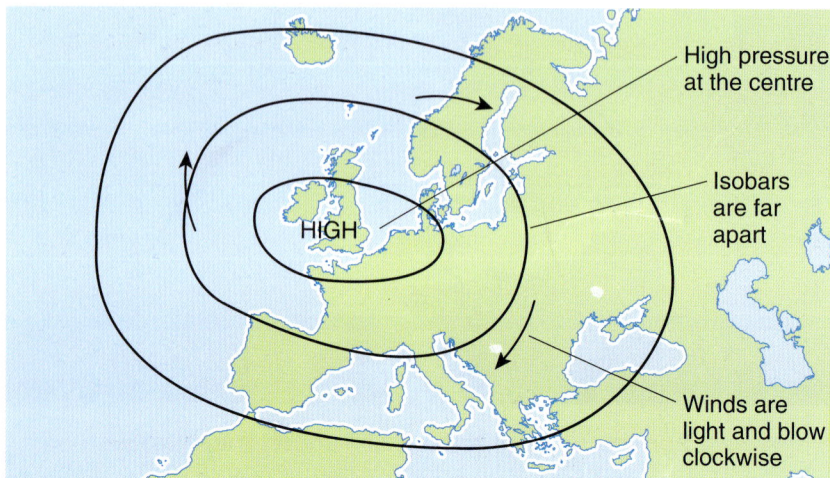

High pressure at the centre

Isobars are far apart

HIGH

Winds are light and blow clockwise

Areas of high pressure are also called ANTICYCLONES. You can recognise an anticyclone on a weather chart because of the word 'high' in the centre. The lines on the chart are ISOBARS. They join places with the same air pressure.

G   An anticyclone on a weather chart

→ ... Low pressure

- Low pressure is when the air is rising from the ground – it often brings wind and rain (photo **H**).

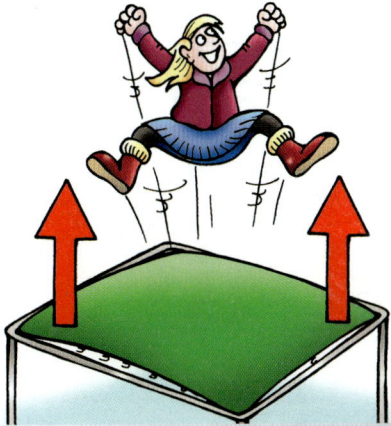

H Camping in low pressure

The air is rising – so it is cloudy

Clouds produce rain

It is sometimes windy too

Clouds keep the heat in, so it is warm at night

Areas of low pressure are called DEPRESSIONS. Notice that the isobars are closer together, so the winds are stronger.

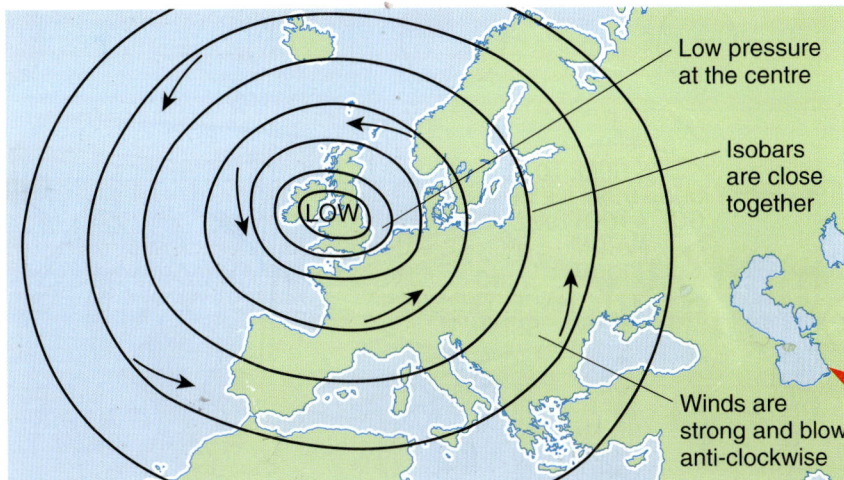

Low pressure at the centre

Isobars are close together

LOW

Winds are strong and blow anti-clockwise

I A depression on a weather chart

➔ A lot can happen in a day

Track a depression

In the British Isles, it's always a good idea to listen to the weather forecast. The weather can change, hour by hour.

A lot of our weather is caused by depressions. They come from the Atlantic Ocean, moving from west to east. First comes the WARM FRONT, then the COLD FRONT (more about them on page 47). Weather forecasters track depressions with the help of satellite images. This helps them to forecast changes in the weather.

The charts below show the progress of a warm front and a cold front as a depression moves across the British Isles. How does the weather change through the day? Read the instruments to record the changes.

07.00 a.m.	09.00 a.m.	11.00 a.m.

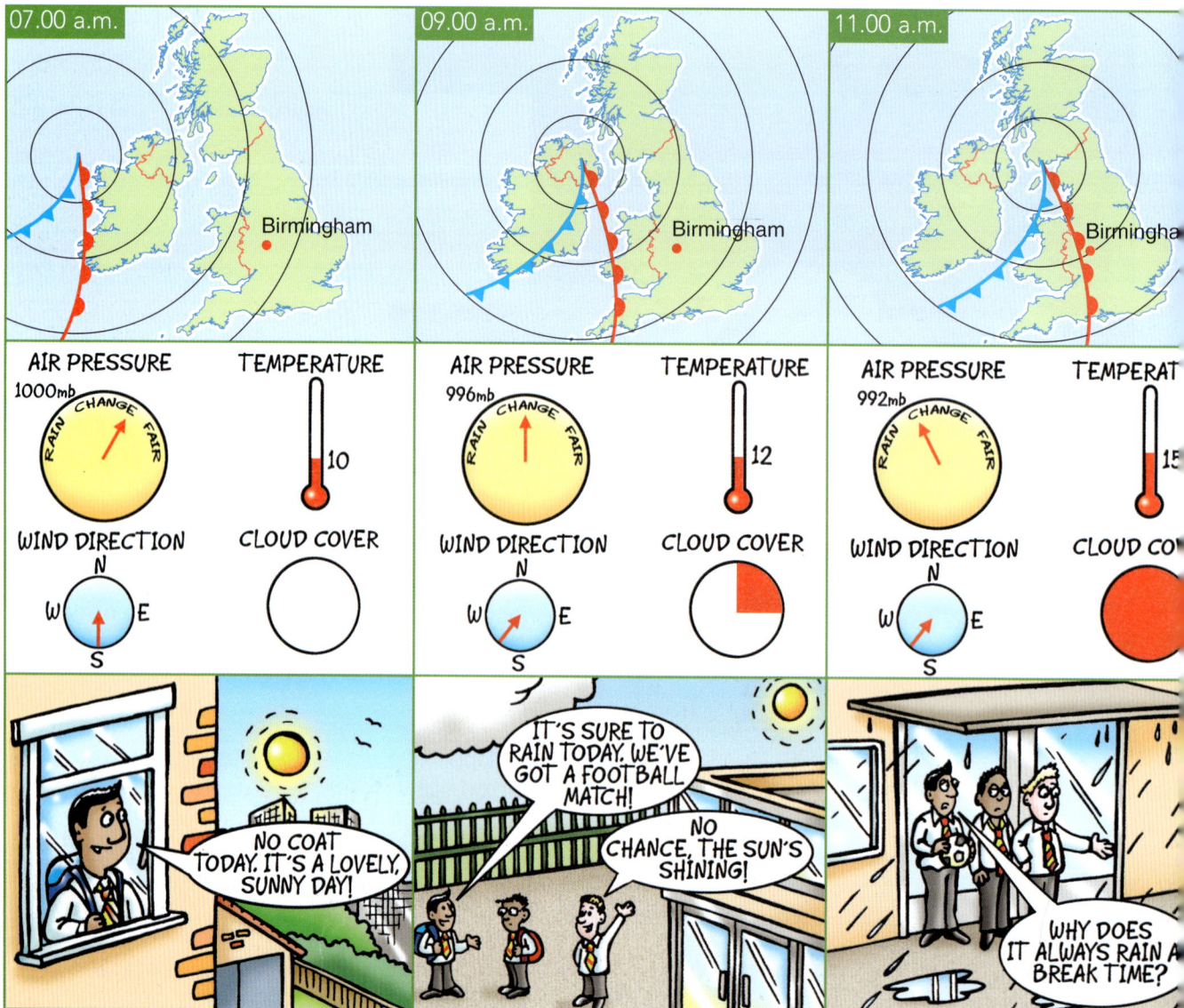

07.00 a.m. — Birmingham

AIR PRESSURE 1000mb — RAIN CHANGE FAIR

TEMPERATURE 10

WIND DIRECTION — N W E S

CLOUD COVER

09.00 a.m. — Birmingham

AIR PRESSURE 996mb — RAIN CHANGE FAIR

TEMPERATURE 12

WIND DIRECTION — N W E S

CLOUD COVER

11.00 a.m. — Birmingha

AIR PRESSURE 992mb — RAIN CHANGE FAIR

TEMPERAT 15

WIND DIRECTION — N W E S

CLOUD CO

NO COAT TODAY. IT'S A LOVELY, SUNNY DAY!

IT'S SURE TO RAIN TODAY. WE'VE GOT A FOOTBALL MATCH!

NO CHANCE, THE SUN'S SHINING!

WHY DOES IT ALWAYS RAIN A BREAK TIME?

activity...

Look at the information on these pages. It records the changes in the weather as a depression passes across the British Isles. Complete a large table like this.

- First, describe the weather in the top row of the table (using the cartoon drawings).
- Then, give the air pressure, temperature, wind direction and cloud cover in the next four rows, using the readings on the instruments.

Time	7.00	9.00	11.00	1.00	3.00	5.00
Weather	sunny					
Air pressure (mb)						
Temperature (°C)						
Wind direction						
Cloud cover (oktas)						

Warm air v cold air

Read a weather chart

In the British Isles weather forecasting is difficult. This is because we live in a battle zone. Cold air blowing from the North Pole meets warm air from the south. The line where they meet is called a front (remember frontal rain on page 39?). Where the warm air rises along the front, it forms depressions.

Weather chart **J** and drawing **K** show a depression over north-west Europe. A warm front is where the warm air rises over the cold air. A cold front is where the cold air pushes below the warm air. The battle is on!

J Weather chart showing a depression over north-west Europe

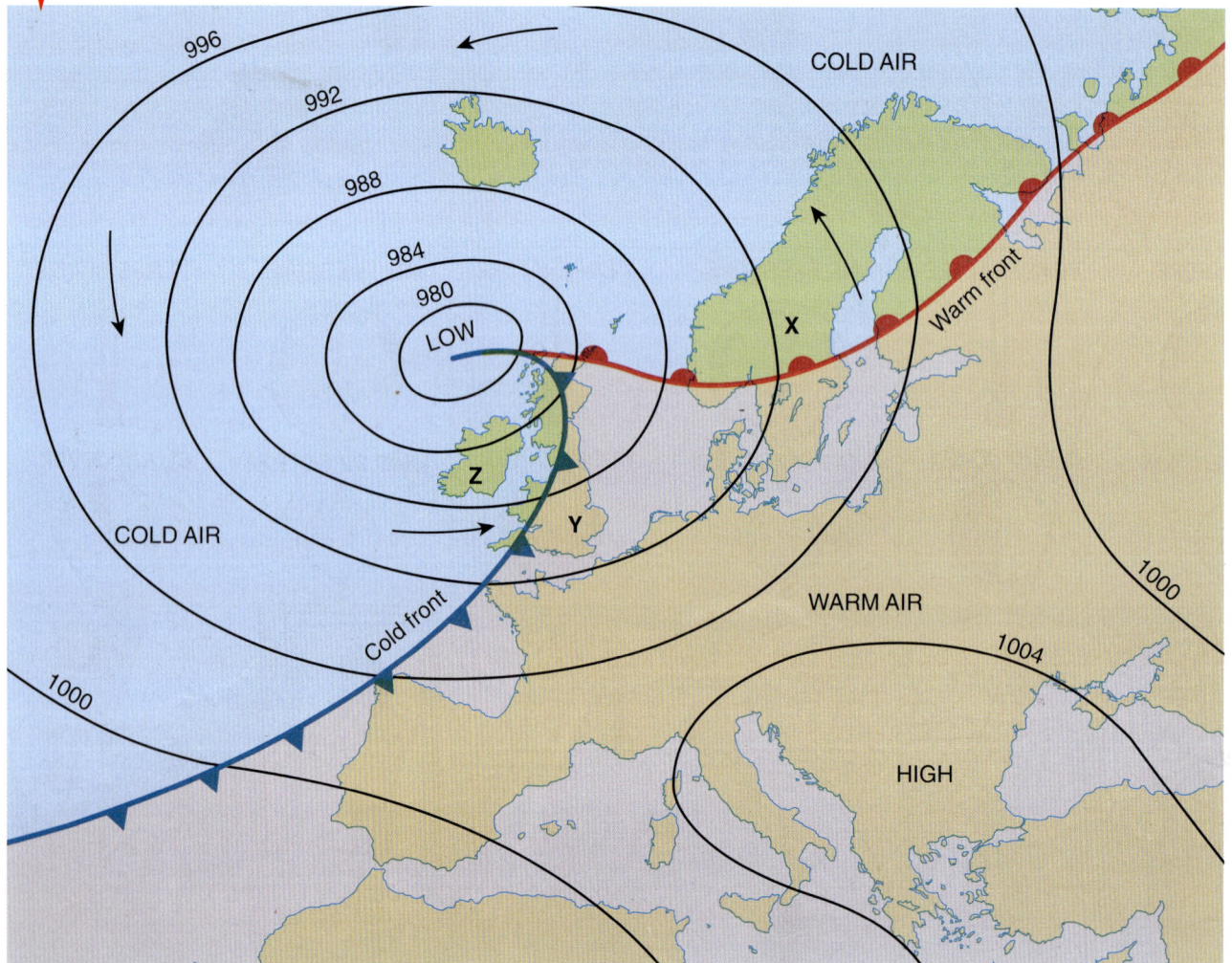

996
992
988
984
980
LOW
COLD AIR
COLD AIR
1000
Cold front
Z
Y
X
Warm front
WARM AIR
1004
1000
HIGH

46

Cumulus cloud is a sign of better weather

Thick cumulonimbus clouds give heavy showers

High cirrus clouds is the first sign of the warm front approaching

Cold front

WARM AIR

Cold air pushes warm air up sharply at the cold front

Layers of stratus cloud bring steady rain

COLD AIR

Warm air slides over cold air

Warm front

COLD AIR

Z   Y   X

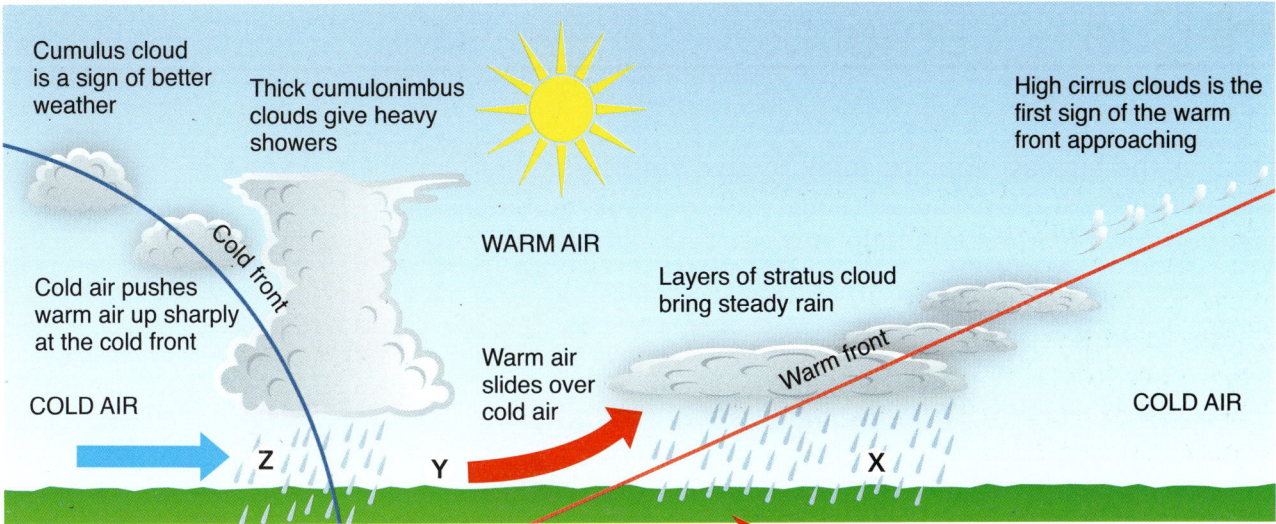

K Cross-section through a depression

activity...

1 Why is it so difficult to forecast the weather in the British Isles?

2 Look at weather chart **J** and drawing **K**.
  a) Name the countries **X**, **Y** and **Z**.
  b) Describe the weather at **X**, **Y** and **Z**. Here are words you can use:

> warm   stratus cloud   steady rain
> sunny   heavy showers   cold   dry
> cumulonimbus cloud

aim high...

3 Look at satellite image **L**. It is the same depression you can see in weather chart **J**. Notice that the cloud pattern follows the fronts on the chart.
  a) Shade the cloud pattern onto an outline map of Europe, using a pencil.
  b) Draw the lines of the warm front and cold front on your map, and label them.
  c) Label where you would find these areas on your map:

> centre of low pressure   warm air
> cold air   cold air

L Satellite image of a depression

47

➔Cue – camera!

Today is the big day. You are going to give your first
weather forecast. Don't get nervous – all the skills and ideas
that you have learned will help you.

■ your final task…

To make your task a bit easier we are going to break it
down into three steps.

Step 1 Make a weather map to show the weather now

1 Look at the satellite image below. It shows a
depression passing over Europe today. Two fronts
have been added to the image. They will help you to
work out where it is raining. What is the weather like
across the British Isles?

2 Draw symbols onto an
outline map of the British
Isles to show the weather
across the country now.
On the right are symbols
that you can use.

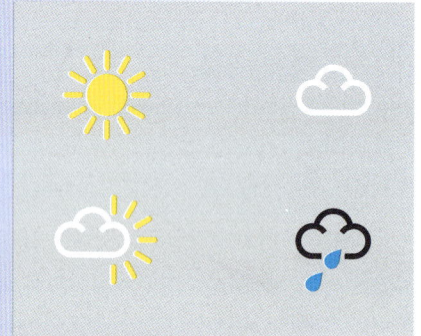

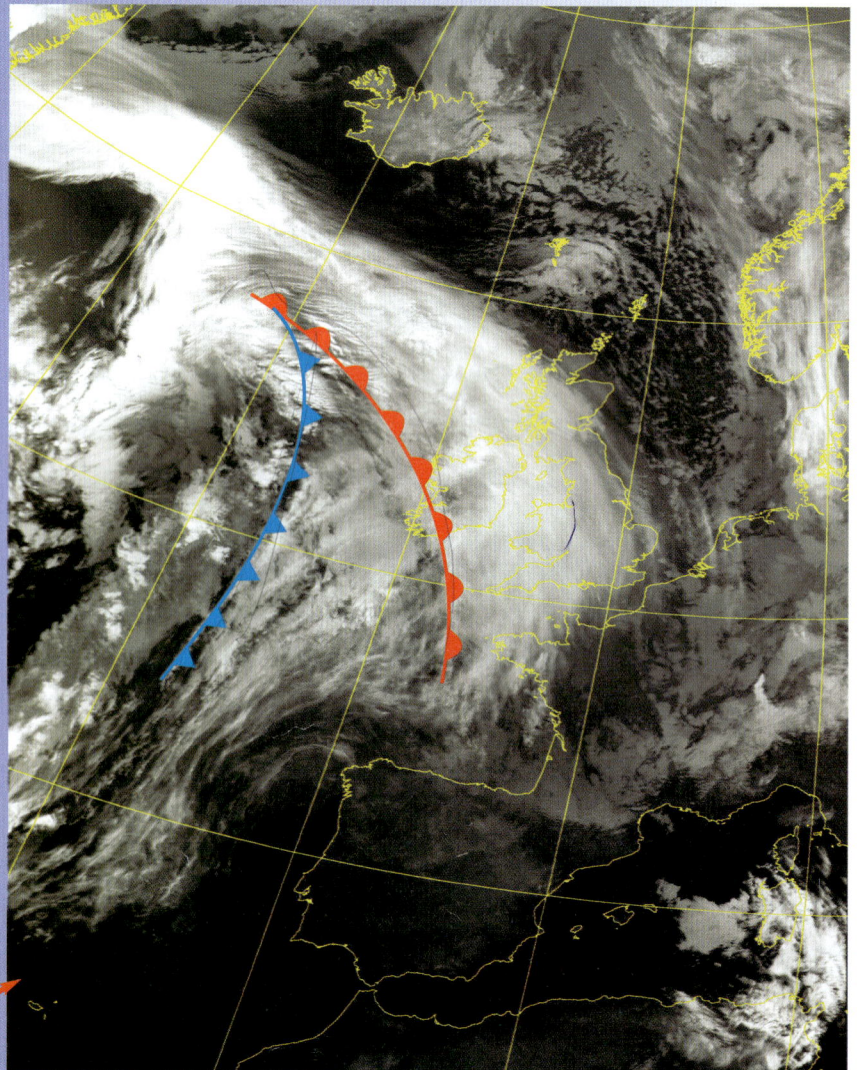

Step 2 Make a weather map to forecast the weather tomorrow

Look at the weather chart. It forecasts the weather tomorrow.

3 Compare the chart with the satellite image. How has the position of the two fronts changed? What will the weather across the British Isles be tomorrow?

4 Draw symbols onto an outline map of the British Isles to forecast the weather in each part of the country tomorrow. Use the same symbols.

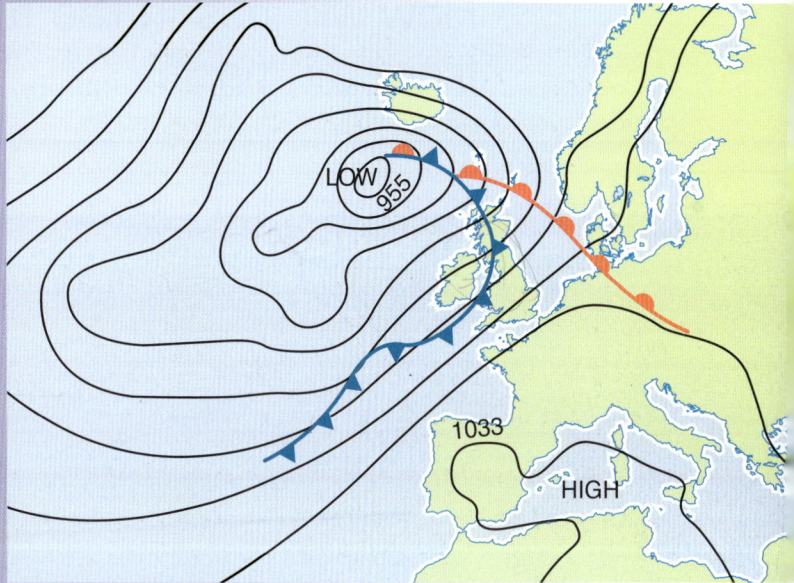

Step 3 Write the script for the weather forecast

5 Write two paragraphs to describe the weather in each part of the British Isles, today and tomorrow.

Your first paragraph could start: 'Today, most of England and Wales is...' Describe the weather on your first map. Write in the present tense.

Your second paragraph could start: 'Tomorrow, the east of the country is likely to be...' Describe the weather on your second map. Write in the future tense (because it is a forecast, you should use words like, could, may be, probably, possibly).

6 Present your weather forecast to the rest of the class. Stand beside a screen, pointing to the maps like a real weather forecaster.

4 Once upon a coal mine

How will jobs in the future be different from the past?

KEY CONCER

- Interdependen
- Place
- Scale

coming up...

The world is changing and so are the jobs we do. This is certainly true in the north-east of England, where the film, *Billy Elliot* was set. Billy's dad was a coal miner and Billy was expected to follow in his footsteps. But, that's not how it turned out ...

your final task...

At the end of the unit you will imagine what the future might bring and produce your own, real-life sequel to *Billy Elliot*.

The Story of Billy Elliot

The date is 1984, the year of the miners' strike in Britain. Coal miners refused to work because the government were closing the mines. Billy is 11 years old. Billy's mum has died and he lives with his dad, his older brother and his gran in a small coal-mining town on the north-east coast of England. One day, during his weekly boxing lesson, Billy stumbles upon a girls' ballet class. He soon discovers that he has more talent for dancing than he does for boxing. But in his village, boys are supposed to be boxers not dancers so, for a while, he keeps the dancing a secret.

Meanwhile, the family is struggling to survive. As the miners' strike drags on for weeks, then months, they run out of money. They have to burn their piano for firewood – because there is no coal! Tensions mount, as one by one, miners go back to work through lack of money.

Then Billy's secret is discovered. His dancing does not go down well with his dad. He wants his son to follow the family tradition and become a miner when he leaves school. Billy has different ideas.

He wants to follow his dream to become a dancer.

Billy Elliot was filmed in and around the mining town of Easington Colliery ('colliery' is another word for a coal mine). In real life, the mine in Easington closed down in 1993, nine years after the miners' strike. Through the rest of this unit you will find out why the mine closed and what has happened in the years since then.

activity...

1 Read the story of *Billy Elliot*.
   a) Why was Billy expected to become a miner?
   b) How might attitudes be different today?
2 Either, look at the photo of Easington Colliery, or watch extracts from the film *Billy Elliot*.
   a) What does it tell you about Easington Colliery in 1984?
   b) How might Easington be different today?
3 Work with a partner. If you have watched the film, think of the good and bad things about living in a small mining town in 1984. Make two lists.

Easington Colliery

51

→ The Cummings family

It is many years since the mine in Easington Colliery closed down. So what are people there doing now?

Some people were lucky enough to find paid work (or EMPLOYMENT) in other jobs. Other people could not find jobs and are still UNEMPLOYED. Whether it is paid work, unpaid work, or leisure, everybody's doing something. Let's catch up with one family in Easington – the Cummings …

6.00 a.m.

Old habits die hard.

You're up early.

Alan Cummings was a coal miner for 30 years. Now, in his fifties, he's unemployed.

8.00 a.m.

Who's my little princess?

At 8 am, granddaughter, Chloe, is dropped o[ff by] her mum. Alan takes her to nursery.

7.30 a.m.

Whose turn is it to cook tonight?

Yours. I'll do the shopping.

Dawn lives with Tony and their children, Chloe and Jack. Tony works as an engineer.

8.00 a.m.

Who are you going to see today?

Grandad

She drops Chloe at her parents on the way t[o] work. Jack looks after himself.

6.00 a.m.

Six o'clock. Time to get up.

Jack is the first in the family to get up. He does a paper round before he goes to school.

11.00 a.m.

Jack Cummin[gs,] Don't go to sle[ep] in my geograp[hy] lesson.

Five hours later, Jack is feeling quite tired i[n] school.

activity...

1 Read the story strips to find out what the Cummings do in a day.
Classify all the activities they do under three headings: 'paid work', 'unpaid work' and 'leisure'. Make three lists.

discuss...

2 When the mine closed, and miners lost their jobs, many of them became depressed.
   a) Why do you think that happened?
   b) How important is it to have a job?
   c) What job do you want to do when you are older? Is that really likely to happen? Why, or why not?

a.m. — I'll dig potatoes up today, and plant a few lettuce.

...nds the morning at his allotment – a plot ...where he grows vegetables.

3.00 p.m. — I can't stop coughing. / It's all those years in the mine. You should get compensation.

Alan still helps the miners' union. He gives advice to other ex-miners.

8.00 p.m. — What's on telly, love?

The end of another busy day. Alan might be retired, but he hasn't stopped working.

a.m. — Hello, Barclaycall. How can I help you?

...works at Barclaycall, a call centre for ...y's Bank. She works from 9 am to 4 pm.

4.30 p.m. — Don't ask! / Busy, Dawn?

On the way home she dashes in to Tesco's to shop. Then picks up Chloe at 5 pm.

7.30 p.m. — Once upon a time ...

After dinner, Dawn puts Chloe to bed. Where has the day gone?

p.m. — Pass inside, Jack.

...chtime there is football training. Jack ...to be a professional footballer.

7.30 p.m. — Don't forget, you're washing up, Jack.

There's no rest, even when he gets home!

8.30 p.m. — Early night, Jack. Remember you've got to get up early.

Just time to do homework before bedtime.

→We were built on coal

Britain is built on coal – literally! Coal is found in rocks below the ground across large areas of the country, called COALFIELDS (map **A**).

For 250 years, from the Industrial Revolution onwards, Britain depended on coal. Towns and villages were built on the coalfields. People came to work in the mines and the factories that grew up around them. Coal provided the power for MANUFACTURING INDUSTRIES like iron and steel, shipbuilding and chemicals.

During the 20th century coal also produced most of our electricity. But the days of coal are almost over. Back in 1963, when Alan started work in Easington, there were nearly a thousand coal mines in Britain. Today there are only a few mines left.

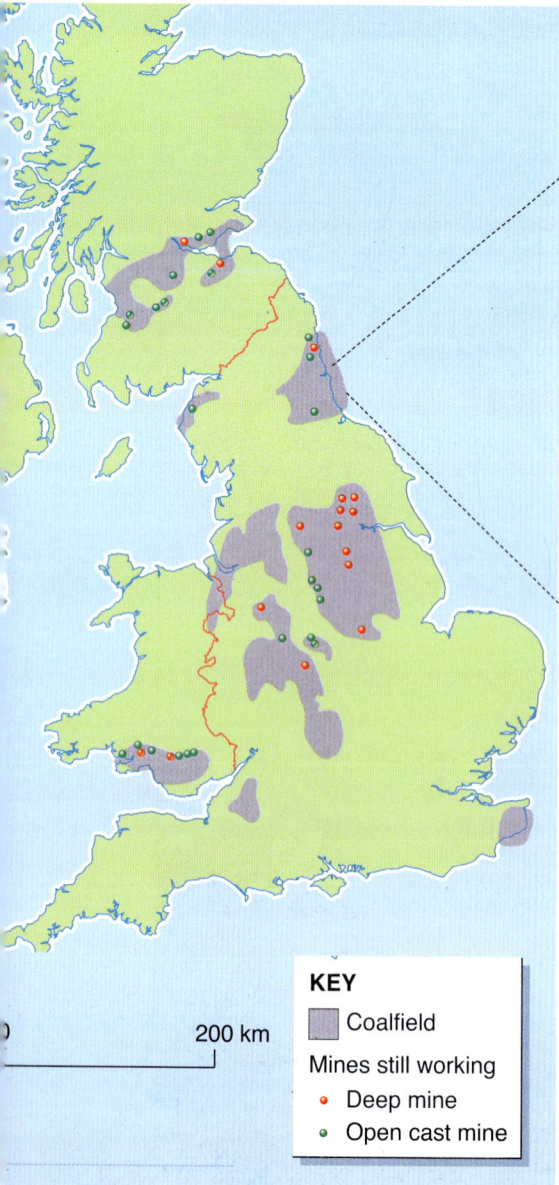

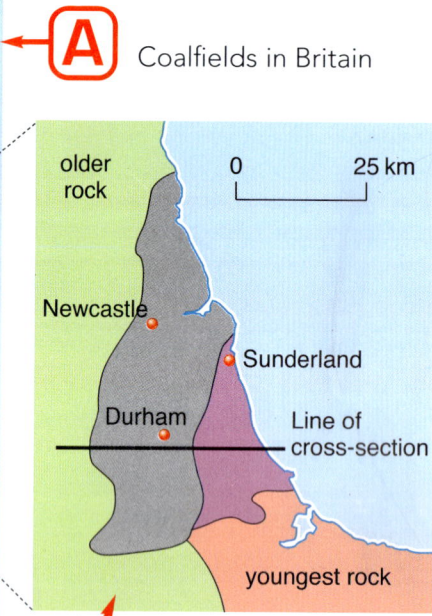

A Coalfields in Britain

KEY
- Coalfield
- Mines still working
  - Deep mine
  - Open cast mine

200 km

B Cross-section of the North-east coalfield

older rock

0   25 km

Newcastle

Sunderland

Durham

Line of cross-section

youngest rock

1700–1800 Coal mining began in the west. The coal here was easy to reach but it ran out in the 19th century.

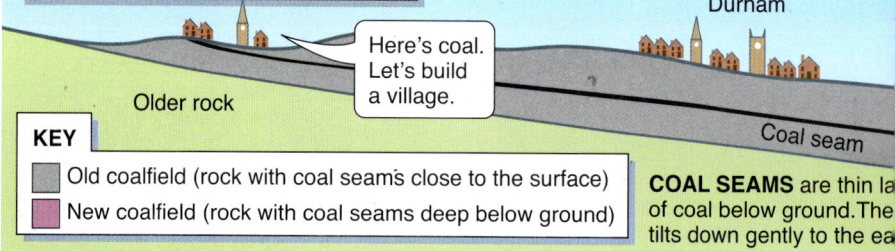

Here's coal. Let's build a village.

Durham

Older rock

Coal seam

KEY
- Old coalfield (rock with coal seams close to the surface)
- New coalfield (rock with coal seams deep below ground)

COAL SEAMS are thin la of coal below ground. The tilts down gently to the ea

activity...

1 You are going to draw a graph to show the decline of coal production in Britain. Use the figures in the table.

a) Draw two axes like this to fill a page in your book.

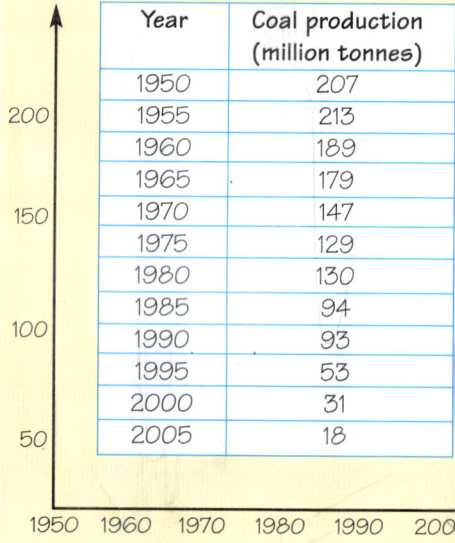

Year	Coal production (million tonnes)
1950	207
1955	213
1960	189
1965	179
1970	147
1975	129
1980	130
1985	94
1990	93
1995	53
2000	31
2005	18

200

150

100

50

1950   1960   1970   1980   1990   200

Alan's story

I was a miner for 30 years. I started work in 1963 when I was 15 years old. My first wage was £18 a week. The work was tough. We worked hundreds of metres underground at the coal face. It was hot and the dust got in your lungs. When I first went down there, we used to use picks and shovels to dig the coal, but later we had machines. Accidents sometimes happened. In 1951, 83 men were killed at Easington. Thank God, I wasn't there.

The number of miners at Easington fell from 2,700 in 1963 to 1,300 in 1993, when the mine closed. Despite this, the mine still produced just as much coal and made a good profit. We were shocked when the government told us that Easington was going to close.

b) Plot the figures in the table on your graph. Join the points with a pencil line.

c) Label the events below on your graph (draw arrows to the correct points on the line):

1956 The first nuclear power station opens
1962 Railways stop using coal for steam power
1965 North Sea gas is discovered. Homes and industry switch from coal to gas
1970 It is getting cheaper to import coal from countries like Poland
1984 National miners' strike. The government import more coal
1990 Power stations start to switch from coal to gas

d) Explain how each event you have labelled could affect the line on the graph.

2 Look at drawing **B**. Write three sentences to describe how mining changed on the North-east coalfield from 1700 to 2000. Start your sentences like this.

Early in the 18th century …
In the 19th century …
By the late 20th century …

aim high...

3 In 1993 the mine at Easington Colliery was closed.
  a) Why did this happen, do you think? The labels on your graph will help you.
  b) Why were the miners shocked when the mine was closed?

Coal's running out at the surface. Let's dig down.

1800–1900 Mine shafts were sunk to reach the coal below ground. The mines got deeper further east.

1900–2000 The deepest mines were near the coast. They went down 1000 metres and extended 10km out, under the sea.

Easington

There's enough coal here to last a lifetime.

North Sea

55

→ Life after coal

Easington Colliery changed forever after the mine closed down. The town used to be full of life. People were not wealthy, but, mostly, they had money in their pockets. Shops and businesses used to thrive.

Today, Easington is one of the poorest areas in the country. Many people are unemployed, while others can't work because of long-term illnesses caused by mining. Shops and businesses have closed. For example, there is no longer a bank in the town.

C Easington High Street today

activity...

1 Look at box **D**. It shows some of the problems in Easington connected with the mine closing. You are going to create a concept map to show these connections.
   a) Write, or draw, the problems on a page in your book, spaced widely apart.
   b) Think about which problems are connected directly with the mine closing. For example, *people leave the area.* Draw lines to connect the problems with the mine closing, like this:

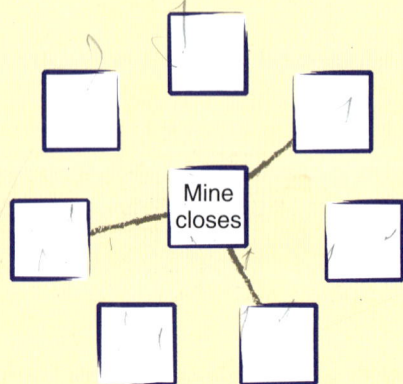

Write labels to explain the connection on each line. For example, *people leave the area when the mine closes to find a job.*

   c) Think about problems that are connected with each other. For example, *people leave the area* and *shops and businesses close.* Draw more lines to connect the problems, like this:

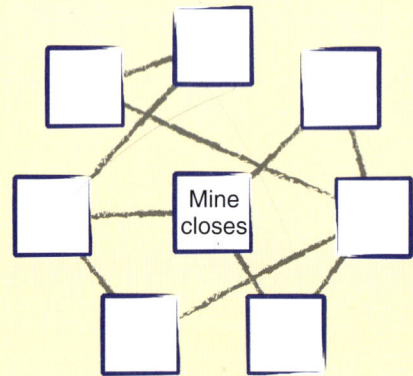

Write labels to explain the connections on each line. For example, *shops and businesses close when people leave because now there are fewer customers.*

discuss...

2 Coal mining was a dirty, dangerous job. Was Easington Colliery better off with, or without, the coal mine, do you think?

I became the local secretary of the union (National Union of Mineworkers) in 1976. My job was to represent the miners in meetings with the mine managers. The union led the miners out on strike in 1984. We were fighting the government to keep the mines open. In the end we lost, and mines in places like Easington eventually closed.

Even today, I spend some of my time working for the union. Ex-miners come to see me with their problems. Lots of them are sick after years of working in the mine. I've got a condition called 'vibration white finger' from holding a drill all day. It prevents me using my fingers.

D Problems caused by the closure of Easington Colliery

PEOPLE LEAVE THE AREA

DAD, I NEED A NEW PAIR OF TRAINERS!

POVERTY

UNEMPLOYMENT

COAL WASTE KEEP OFF

SPOILT LANDSCAPE

MINE CLOSES

PIT CLOSED

MR. JONES TO DR JOHN ROOM 8 PLEASE

POOR HEALTH

CRIME AND VANDALISM

GCSE GEOGRAPHY

MARK 20%

FAIL

POOR EDUCATION RESULTS

NAT-EASTERN BANK   TRAVEL Co.

CLOSED!   GONE AWAY!

SHOPS AND BUSINESSES CLOSE

→ Calling new industries

Traditional industries, like coal mining and shipbuilding have almost disappeared from the North-east. However, new industries, like car manufacturing, electronics, bio-technology and tourism have come to take their place. The fastest-growing industries in the region are SERVICES, like the new call centres. They now employ 40,000 people in the North-east alone. They deal with customer phone calls for big organisations like banks or insurance companies.

Doxford International Business Park, near Sunderland, is one of the largest call centre developments in the UK.

E Doxford International Business Park

The North-east has a workforce of over one million people living within easy travelling distance of Doxford. Six per cent of the workforce is unemployed – more than other parts of the UK.

Doxford was built at the junction of the A19 and A690. It is close to the A1(M) motorway and to Newcastle International Airport. It has good access to Europe and other parts of the UK.

The business park was built outside the city, on green land, where rents are lower. There is plenty of space for car parking and for future expansion. Companies at Doxford include: Barclays, The National Lottery, Nike and T-Mobile.

About 8,000 people work in call centres at Doxford. Three-quarters of them are women. Wages here are 30% less than in the South-east.

The buildings are modern and Doxford has the latest telecommunications systems. It has one of the few earth satellite stations in the UK, making calls cheaper.

Dawn's story

Working at Barclaycall is convenient. It takes me only fifteen minutes to drive to Doxford in my car. And the hours are flexible, so it fits in with family life.

I didn't need any bank experience to get the job – just common sense and a nice friendly voice. I had to do four weeks' training before I started.

Of course, the money helps. The wages are not fantastic – about £15,000 a year. That's £300 a week for 35 hours work. It's more than I'd get working in a shop. The main problem is the stress. Dealing with difficult customers on the phone all day can wear you down. I get tired and irritable.

activity...

1 You work for One North-east (ONE), the agency whose job is to attract companies to the North-east.

  a) Look at photo **E**. List the main advantages that Doxford International Business Park could offer a company moving to the North-east.

  b) Design a page for the ONE website to attract companies to Doxford.

2 Look at this pie chart. It shows the economic activities that people did in the North-east in 1950.

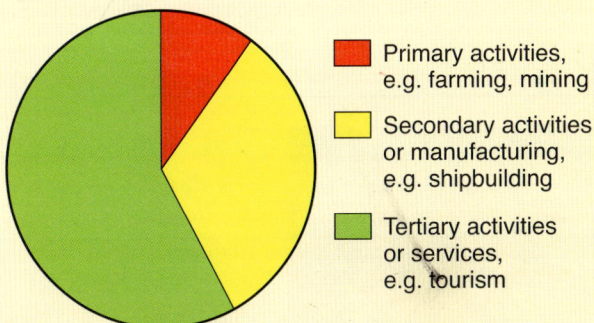

- 🟥 Primary activities, e.g. farming, mining
- 🟨 Secondary activities or manufacturing, e.g. shipbuilding
- 🟩 Tertiary activities or services, e.g. tourism

  a) Draw a similar pie chart to show the activities that people in the North-east do today, using the figures below:

Economic activity	Percentage
Primary	2%
Secondary	22%
Tertiary	76%

  b) Compare the pie chart you have drawn with the one above. Describe how economic activities in the North-east have changed since 1950.

  c) How do you explain these changes from what you have learnt in this unit?

discuss...

3 **a)** How do you think the changes you described in Activity **2** would affect the number of jobs for men and women?

  b) What impact would these employment changes have on family life?

  c) Would these be good or bad changes, do you think?

➔ Here today, gone tomorrow?

Many of the companies that have come to the North-east are TRANSNATIONAL COMPANIES, or TNCs – big companies based in one country, but with branches all over the world. This has happened because of GLOBALISATION. Globalisation is the way that jobs, people and ideas now move around the world.

The Japanese car-manufacturing company, Nissan, set up a new factory near Sunderland in 1986. It now employs 4,300 people and produces 300,000 cars every year.

Why transnational?
In the 1960s and 70s, Japanese companies, like Nissan, were very successful. They produced better cars more cheaply than companies elsewhere.

> We can't sell any more cars in Japan. Where can we go now?

Why Europe?

> There are half a billion people living here. That's a lot of cars!

> The European Union puts tax on cars made abroad. We need to build our factory in Europe.

Why the North-east?

> The workforce here have engineering skills. They used to build ships.

> There is a big port nearby. So we can send our cars all over Europe.

Why the UK?

> I've heard the UK government gives generous grants to bring in new industries.

> Wages in the UK are lower than some Euro countries. That would keep costs down.

Why this site?

> An old airfield – the perfect site. A huge flat area where we can build a factory quickly.

> And a major road right on our doorstep!

But, for how long?
The Nissan factory near Sunderland today. But there have been rumours that the company wants to move.

The problem with TNCs is that they come – but they can also go. Take the German electonics company, Siemens. It set up a new factory in Newcastle in 1995. The factory closed in 1998 with the loss of 1,100 jobs. Companies that set up call centres in the North-east are now moving abroad where the wages are lower.

(F)

Extract adapted from the *Guardian* newspaper

AT YOUR SERVICE: INDIANS HEED CALL OF THE WEST

Charles Haviland in Bangalore

Cyril Deepak spends his time at work on the phone to Britain. He is learning to answer customer calls for a large British insurance company. Deepak, 24, is one of over 12,000 call centre workers in the Indian city of Bangalore employed by British or American companies.

They will soon be joined by others. The National Rail Enquiry service says it wants to move to India to cut costs. HSBC Bank has just announced it is cutting 4,000 call centre jobs in Britain and moving them to Asia. BT, British Airways, Lloyd's TSB, Prudential, Norwich Union, BUPA, Abbey National and Powergen have already begun to move their call centres to India.

It is not hard to see why most of them have chosen India. The wages of workers in the service industries are one-tenth of workers over here. The starting salary for an Indian call centre worker is between £130 and £200 a month. Standards of education are high and almost all educated Indians speak English. While British workers will take call centre jobs only when they have no choice, Indian workers see them as glamorous. One company in Bangalore recently advertised 800 jobs. It received 87,000 applications.

activity...

1 Study the information on page 60. Why did Nissan choose this site for their factory in the North-east?

a) Draw five circles to fill a page in your book, like this.

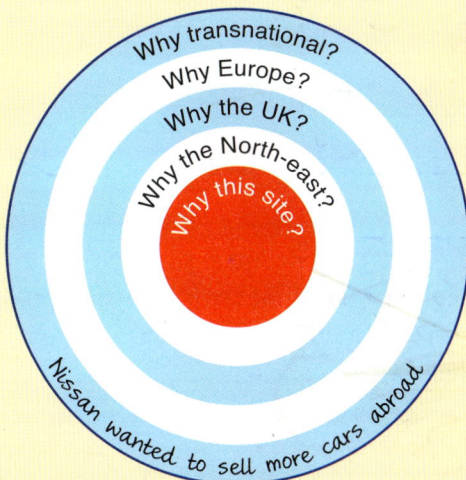

Why transnational?
Why Europe?
Why the UK?
Why the North-east?
Why this site?

Nissan wanted to sell more cars abroad

b) List the reasons that Nissan came here under the headings in each circle. Start from the furthest circle from the centre. One is done for you.

2 Read newspaper article **F**.
What views would these people have about globalisation? In each case, say if they would be For, Against or Don't mind. Write a sentence to give a reason.

a) A call centre worker in North-east England.
b) A call centre worker in Bangalore.
c) A director of HSBC Bank.
d) A telephone customer of HSBC Bank.
e) The British Prime Minister.
f) You!

aim high...

3 Imagine a discussion between Alan Cummings and his daughter, Dawn. Alan thinks that transnational companies are bad for the North-east, but Dawn thinks they are good. Who do you agree with?

a) Think of at least three arguments to support your view.
b) Find someone in your class with the opposite view. Now, role-play the discussion that Alan and Dawn might have.

→ What will Jack do?

How often do adults ask you 'What will you do when you leave school?' What a hard question – especially when jobs change so fast these days!

The problem with some adults is that they seem to think that the world hasn't changed since they were young. But, if you've studied this unit, you'll know it has.

Lots of adults are keen to give Jack advice – but who should he listen to? Or should he be like Billy Elliot and follow his dream to be a footballer?

> In my day we left school at 15 and went down the mine. My advice is, when you leave school, go out and find a good trade. People will always need plumbers!

Grandad

> You might not be good enough to become a professional footballer. Whatever you do, don't fall behind with your schoolwork. These days everybody needs qualifications.

Mum

Employer

> Our company offers a wide range of careers for well-qualified school leavers. When you come to us at 16 we will train you and you could be earning £20,000 a year by the time you are 20.

Teacher

> You've got the potential to go to university, Jack. It doesn't matter if your friends all go to work in McDonald's. People around here are not ambitious enough.

discuss...

1 Read the advice that Jack has been given. Is it good advice or not?

2 Think about what you have learnt in this unit. What advice would you give Jack about what to do when he leaves school?

■ your final task...

You have been asked to produce ideas for a new film set in the north-east of England – a sort of sequel to *Billy Elliot*. This film is going to be about what happens to Jack.

This is what to do.

1 Think about what could happen to Jack. Turn it into a good story. Don't forget what you have learnt in this unit (this will give you some ideas), but use some imagination too. Below are two alternative ideas to get you started. Or, think of your own idea.

2 Divide the story into scenes. For each scene, think of a location, e.g. outside the school gate, and what happens, e.g. Jack tells his friends he is leaving school now he is 16.

3 Draw a storyboard to illustrate each scene, like this. Keep the drawings very simple. Your film should have at least ten scenes. You *don't* have to write the script for the film.

Scenario 1

Location: Outside the school gate
What happens: Jack tells his friends he is leaving school now he is 16

Location: In the office of a large company
What happens: Jack attends an interview for a job as a trainee manager

?

Scenario 2

Location: Sitting in a school classroom
What happens: Jack stays at school and is studying for A Levels

Location: In the kitchen at home
What happens: Jack opens the envelope with his A Level results

?

A question of football
What's football got to do with geography?

KEY CONCEP

- **Scale**
- Space
- Cultural understanding and diversity

▌▌ coming up...

Did you know that football and geography have a lot in common? I wonder if your teacher agrees? This unit will help you to prove it.

▌▌ through the unit...

You will ask lots of questions to explore different ways in which football is connected with geography. Asking questions is one of the best ways to learn more about geography. (You should try it more often!)

◼ your final task...

At the end of the unit, you will try to persuade your teacher that you really should spend more time studying football in geography. To do that you will need to think of a good geographical question. Then you might get extra time to investigate it!

Look at this football stadium. So many people ... and so many questions you could ask!

What is the weather?

Where is the stadium?

What impact does the stadium have on the surrounding area?

How do they keep the grass so green?

Which countrie do the players come from?

Should the stadium have been built here?

How many people are there in the crowd?

What types of people go to football matches?

activity...

1 Look at the questions around the photo.

  a) Which questions do you think are geographical (have something to do with geography)? Make a list.

  b) Which do you think is the most interesting geographical question? Explain why you think it is interesting.

2 Think about how you would answer all the questions. (You don't need to know the answer.) Classify the questions into two groups:

  • questions that only need a one or two word answer (these are closed questions)

  • questions that need a longer answer (these are open questions).

How much did they pay to get in?

Can everyone afford a ticket?

Which teams are playing?

How far have they travelled?

Why does English football attract foreign players?

Why are these teams so successful?

How did the spectators all get here?

Where will people watch the match on TV?

How much do the players earn?

Do players deserve so much money?

➜ What is a geographical question?

You could argue that all the questions on pages 64 to 65 are geographical. But what is a geographical question?

There are six types of question that we often ask in geography. What questions would you ask about photo **A**?

A

WHERE? To find out where a place is.

WHAT? To find out what you can see in a picture, or what is happening.

WHY? To find out why it is there, or why it happens.

HOW? To find out how it happens.

WHAT IMPACT? To find out how it can affect people, or other places.

WHAT MIGHT? or **WHAT SHOULD?** To find out what might happen, or what should happen, in the future.

activity...

1 Think of a *WHERE?*, a *WHAT?*, a *WHY?* and a *HOW?* question to ask about photo **A**.

aim high...

2 Think of a *WHAT IMPACT?* and a *WHAT MIGHT/WHAT SHOULD?* question to ask about photo **A**. These are more difficult questions, but they can also be the most interesting.

The best geographical questions are not always the most obvious ones. For example, lots of football supporters often ask, Why does my team never win? Photo **B** shows you how this could be a geographical question.

B Why does my football team never win?

Not many people live here, so there's not many players to choose from.

It's a long way from anywhere else, so the team doesn't get much practice against other teams.

There's no stadium, so the team doesn't get many supporters.

Without many supporters there's not much money to buy better players.

Here are three simple rules to help you to think of a good geographical question:
- **Rule 1** A question that geography will help you to answer.
- **Rule 2** An open question that needs more than a one-word, or two-word, answer.
- **Rule 3** A question that you REALLY want to know the answer to.

activity...

3 Go back to the photo on pages 64 to 65.
   a) Choose one question that you think is a good geographical question. It might be the same question that you thought was most interesting.
   b) Explain why it is a good geographical question.

aim high...

4 Do some geographical research into your favourite football team. If you don't have a favourite, choose a local team. A good place to do your research would be their website on the internet. Think of some geographical questions that you could ask. Here are some questions to get you started (add more of your own):

Where do the players come from?
What is the furthest the team has to travel to an away game?
How do you get to the stadium?

→Off to the match

Being a football supporter is a good test of your geography. If you go to your team's away games you have to travel all over the country. And that means using maps!

Imagine that your team has been drawn away to Birmingham City in the next round of the FA Cup.

C Motorways and major roads in England

activity...

1 Look at map **C**. Find where your team is on the map. If you live in Birmingham you can choose another team.

 a) Work out the most direct route to Birmingham from where your team is.

 b) Measure the distance by road. Use a strip of paper or thread to follow the route on the map. Then straighten the strip and place it on the scale to find the distance.

 c) Work out the time your route would take if your average speed is 100 km per hour. This is how you do it:

$$\frac{distance\ (km)}{100\ (km/hour)} = time\ (hours)$$

 d) These days, many people use the internet to plan their route. Try this route planner: www.rac.co.uk/web/routeplanner/ Compare the route to Birmingham it suggests to the one you planned. Which is the shortest and quickest?

Newcastle 3.50
Carlisle 2.50
Durham
Darlington
Middlesbrough
Barrow
Lancaster
Blackpool
Preston 1.40
Bradford
Scarborough
Bolton
Leeds
York 2.15
Manchester
Wakefield
Hull 3.00
Liverpool 1.35
Doncaster
Grimsby
Cleethorpes
Llandudno
Stockport
Scunthorpe
Holyhead 3.20
Stoke
Lincoln
Bangor
Chester
0.50
Sheffield
Crewe
1.10
Newark
Shrewsbury
Telford
Derby 0.40
Nottingham 1.20
Tamworth
Burton
Aberystwyth 2.45
Wolverhampton
Nuneaton
Peterborough
0.15
Norwich 3.25
BIRMINGHAM
Leicester 1.00
Worcester
Coventry
Rugby Northampton
Cambridge
Hereford
0.25
Milton
Keynes
Fishguard 4.35
Banbury
High
Watford
Cheltenham
Wycombe
Euston 1.40
Carmarthen
Gloucester
Oxford
Cardiff 1.55
Marylebone 2.20
Swansea 2.55
Reading
Newport
Didcot
Paddington 2.10
Bristol 1.30
Bath
Basingstoke
Salisbury
Taunton
Penzance
Brighton 3.20
Plymouth Exeter
Southampton Portsmouth
3.45   2.40
2.40   2.40

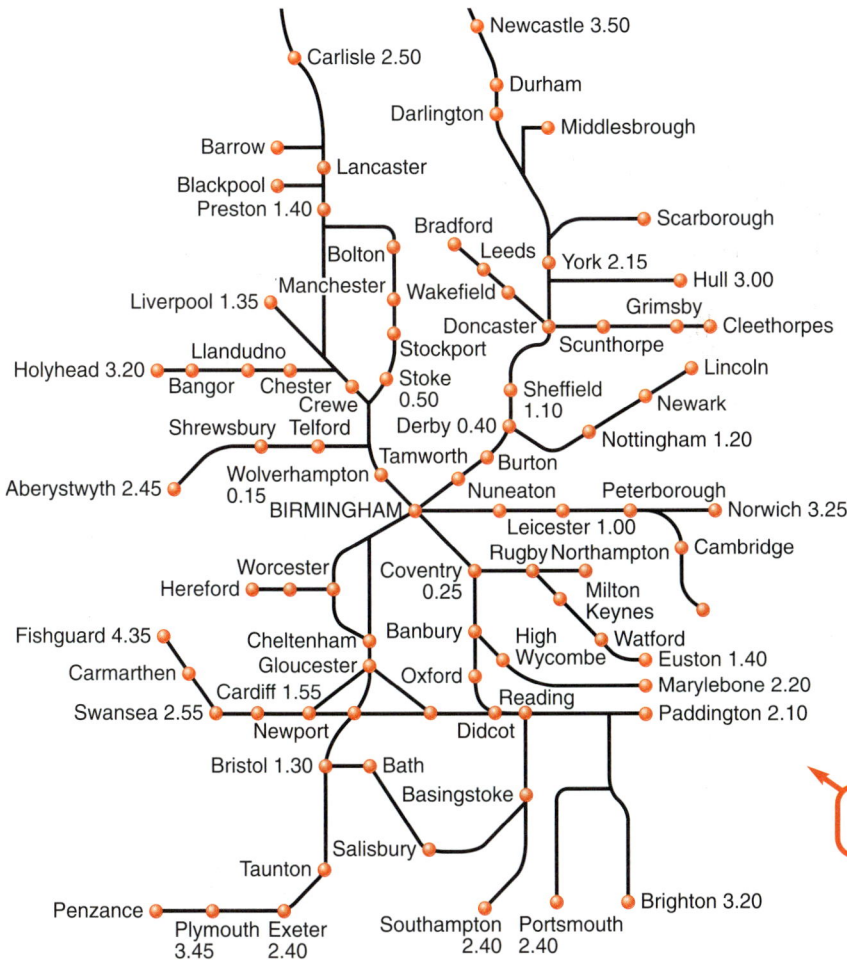

D Train routes and times to Birmingham. For example, 1.10 means the journey takes 1 hour 10 minutes.

activity...

2 Now, work out the time your journey would take by train.
   a) Find your nearest station on map **D**. How long is the journey to Birmingham? (NB Times are given in hours and minutes, e.g. Brighton 3.20, but you will need to estimate times from some stations.)
   b) Add the time it would take you to get to the station.
   c) When you arrive in Birmingham it would take another 30 minutes to walk from the station to the stadium. Work out the total journey time by train, including getting to and from the stations.

aim high...

3 Work with a partner. You are going to travel together to Birmingham. Decide whether to travel by car or train. Here are some of the factors you need to consider:
   • journey time (from Activities **1** and **2**)
   • convenience
   • cost (petrol costs about 5p per kilometre); you can find train fares at www.nationalrail.co.uk (don't forget there are two of you!)
   • possible delays (traffic jams or train cancellations).

any questions...

Think of some geographical questions that you could ask now you have studied these pages. Think of a WHERE?, a WHAT?, a WHY? and a HOW? question. For example:
What is the best route to Birmingham from here?

→Football for all

Did you think that football was just a boy's game? Think again.

Way back in 1920, just after the First World War, 53,000 people came to watch a women's football match at Everton FC's stadium, Goodison Park. At the time, women's football was becoming popular. But the Football Association (the FA) received complaints about women playing football and decided to ban it.

Now, over 80 years later, women's football is making a comeback. It is the fastest growing sport in the country and this time it is supported by the FA. They run the Women's Premier League in England (**F**). Who says football is just a boy's game?

E A women's match

Position		Points
1	Arsenal	62
2	Everton	57
3	Leeds	40
4	Bristol	34
5	Chelsea	32
6	Doncaster	29
7	Watford	29
8	Blackburn	28
9	Birmingham	25
10	Liverpool	22
11	Cardiff *	12
12	Charlton *	4

* Relegated teams

F Women's teams in the FA Premier League in England, 2007–8

activity...

1 Find the latest (men's) Premier League table. Compare it with the women's Premier League in **F**.
   a) Which clubs have teams in both leagues?
   b Which clubs only have a team in the women's Premier League?
   c) Which well-known clubs in the men's Premier League are missing?
2 Look at table **H**.
   a) Put the countries into rank order, according to the proportion of women footballers.
   b) Which three countries have the highest per cent of women footballers? Where are they?
   c) Which three countries have the lowest per cent of women footballers? Where are they?
   d) How does England compare to other countries?

Around the world, women are still breaking down barriers in the traditional men's game (article **G**). Women's football in England still has a way to go before it is as popular as in some other countries (table **H**).

G Article adapted from the *Guardian* newspaper (18th December 2004)

Mexican football club signs woman

JO TUCKMANN IN MEXICO CITY

A Mexican professional football club has made history by signing a woman. The move has already caused uproar in Mexico but the player is not worried. 'I'm not frightened of anything,' Maribel Dominguez told reporters at a packed press conference called by the second division club, Celaya.

'I want to thank all those who believed in me and ask those who don't to give me the chance to try. Maybe I will fail, but at least I will have tried.' Player and club insist there is nothing in the rules to prevent women from playing in the men's professional league.

Generally accepted as the best female footballer in Mexico, Dominguez scored 45 goals in 46 matches for the women's national side. Her squad reached the quarter-finals at the Athens Olympics, while the men's team was knocked out in the first round.

The 26-year-old striker's determination to play is already sending shockwaves through Mexico. But Dominguez said she was ready for the physical strains and the taunts she expects from fans. She recalled how her primary school headmaster once threatened her with expulsion if she

continued playing football with boys. 'I didn't pay any attention then,' she said.

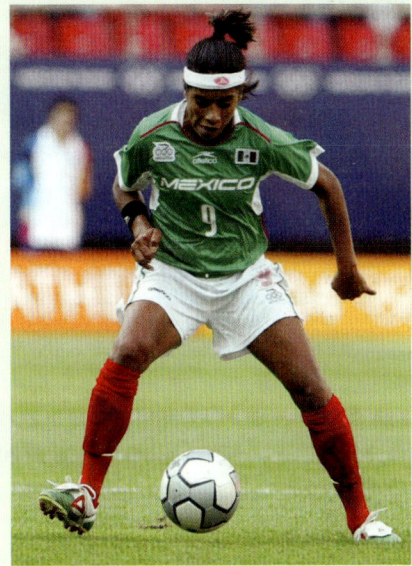

discuss...

3 Read newspaper article **G**. Then, discuss these questions with a partner.
   a) Why don't girls always get the same opportunities to play football as boys?
   b) Should women play football in the same league as men? Why or why not?
   c) Is women's football in England ever likely to be as popular as men's football? Why or why not?

any questions...

Think of some geographical questions that you could ask now you have studied these pages. Think of a *WHERE?*, a *WHAT?*, a *WHY?* and a *HOW?* question. For example:
Where can girls play football in our area?

H Percentage of women amongst footballers worldwide

Country	Women footballers (%)
Argentina	3.1
Australia	16.1
Brazil	2.3
Cameroon	6.2
England	2.3
France	1.7
Germany	13.9
Indonesia	4.5
Italy	0.6
Mexico	6.7
Nigeria	3.5
Netherlands	17.0
Norway	16.7
Spain	0.3
Sweden	25.0

Source: The First World Atlas of Football

➜ Home …

Over the past few years, many football clubs have moved to a new stadium out of town. One club has gone a step further. It has moved to another town altogether!

In 2003, Wimbledon Football Club left south London, its home for the past hundred years, and moved to Milton Keynes. With only a few thousand supporters, and without a permanent stadium of their own, the club decided it was time to leave. Milton Keynes is the fastest growing town in the UK, with 300,000 people – and, until Wimbledon arrived, it did not have a professional team. Now, the club is called MK Dons.

Wimbledon's former stadium at Plough Lane in south-west London

Come on Wimbledon! The club has been torn from its South London roots. How can you move a team and its supporters 100 km from their true home? The directors of the club are only interested in money. That's why they have gone to Milton Keynes.

Real Dons supporters have started a new team – AFC Wimbledon. We're back where we were 30 years ago – in the Isthmian League. But, it won't be long before we catch up the MK Dons and overtake them.

An AFC Wimbledon supporter

activity…

1 Work with a partner.
You have to assess Wimbledon's decision to move to Milton Keynes. Was it the right decision, or should they have stayed in London?
a) Individually, study all the information on these pages, including the maps and photos.
b) Then, together, complete a large table, like this, to compare the two alternatives. Give each alternative a score out of 5 for each factor (5 if it's perfect, 0 if it's awful). Work out a total score for each alternative.
c) Finally, assess if the club made the right decision.

	Stay in Wimbledon	Move to Milton Keynes
Local support		
Good transport links		
Space to build a big stadium		
Cost of land and building (out of town is cheaper)		
Impact on neighbours (especially traffic/ noise)		
TOTAL SCORE (out of 25)		

→ ... Or away

K

J The site of the new Dons stadium in Milton Keynes

Up the Dons! Milton Keynes is a new town. Most of the people living here have moved from somewhere else – many of them from London. Why shouldn't a football team move here too? We desperately need our own team to support.

We offered Wimbledon a site to build a new, state-of-the-art stadium for 30,000 people. With all the support here, MK Dons will soon be back in the Premier League.

An MK Dons supporter

aim high...

2 In the season 2006 to 2007, MK Dons were in League One of the Football League and AFC Wimbledon were in Division One of the Isthmian League (four divisions below). Now, do some internet research to answer these questions:

a) Find out which league the two teams are in now. Have their positions changed?

b) Which club made the best decision, do you think?

c) Which other clubs have a new stadium?

any questions...

Think of some geographical questions that you could ask now you have studied these pages. Think of a *WHERE?*, a *WHAT?*, a *WHY?* and a *HOW?* question. For example:

Why has our local team moved to a new stadium?

73

➡European Champion's League

The Champion's League is the most important club competition in Europe. The top teams in each country compete against each other. Nearly all the teams come from big cities with large populations. So, do the biggest cities have the most successful teams? You are going to investigate.

activity...

1 Look at map **L**. Compare the location of the top sixteen teams in the Champion's League with population density. What do you notice?

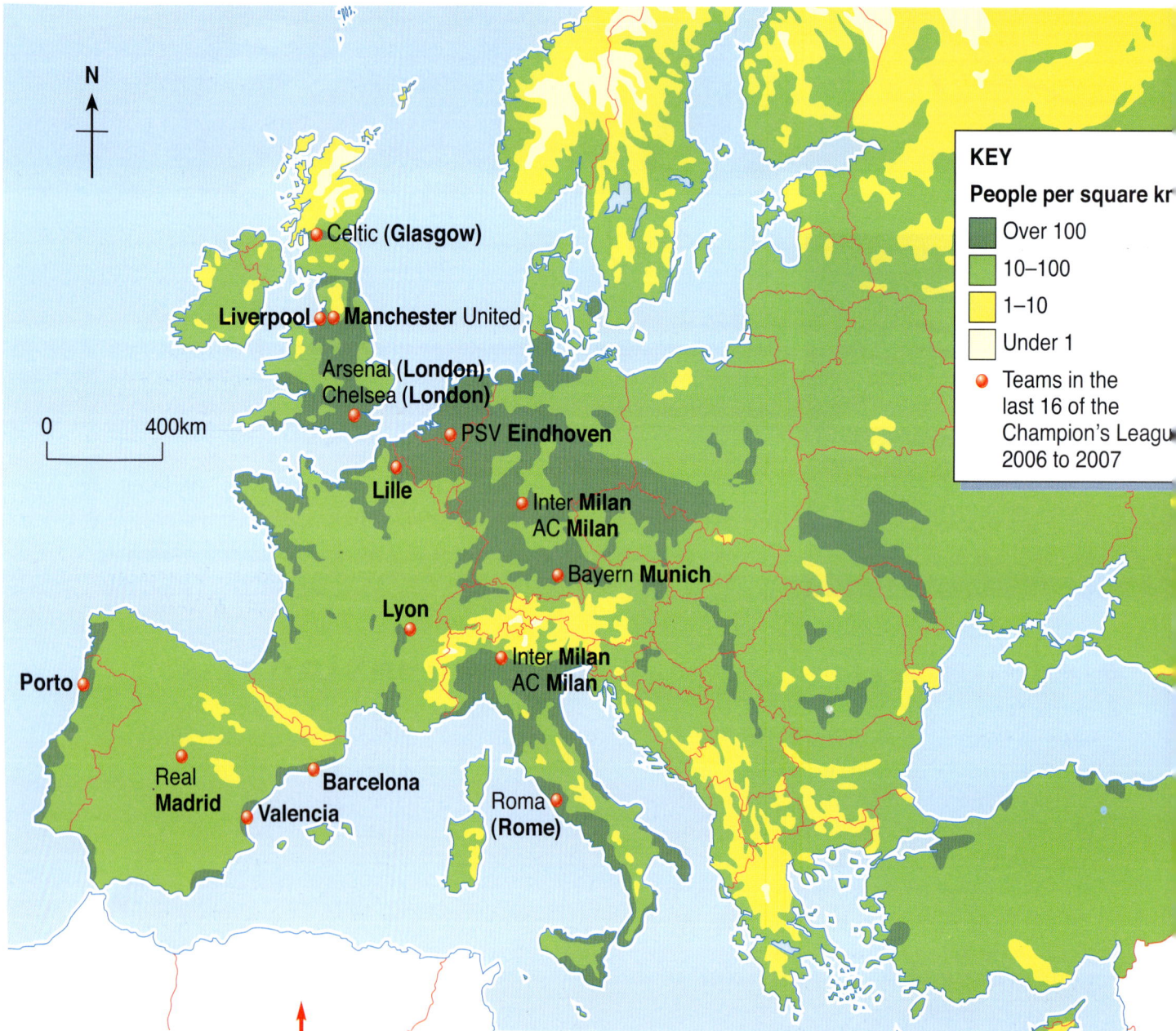

KEY

People per square km

- Over 100
- 10–100
- 1–10
- Under 1
- Teams in the last 16 of the Champion's League 2006 to 2007

Map labels: Celtic (**Glasgow**), **Liverpool**, **Manchester** United, Arsenal (**London**), Chelsea (**London**), PSV **Eindhoven**, **Lille**, Inter **Milan** AC **Milan**, Bayern **Munich**, **Lyon**, Inter **Milan** AC **Milan**, **Porto**, Real **Madrid**, **Valencia**, **Barcelona**, Roma (**Rome**)

Scale: 0 – 400km

L Population density in Europe

2 Now, you are going to look more closely at the connection between population and football success. You are going to test this hypothesis (or theory):

The bigger a city, the more successful its football team.

a) Complete a copy of table **M**. Rank the teams in order of the city's population, from 1 to 16. The largest city is ranked 1, the smallest city 16. Two teams in the same city get the same rank then skip a rank before the next team.

b) Make a large copy of the grid below. Plot the teams on the graph using their population rank and the stage they reached in the Champion's League. One is done for you. You could update the final column to the past season if you wish.

c) Look at the pattern on the graph. Can you see any link between population and the clubs' position in the Champion's League? Describe the pattern that you can see.

d) Write a sentence to say if you have proved or disproved the hypothesis.

aim high...

3 a) Explain the connection, or lack of it, between population and a football team's success. (Photo **B** on page 67 may also help).

b) What other factors could help to make a team successful? Write a list.

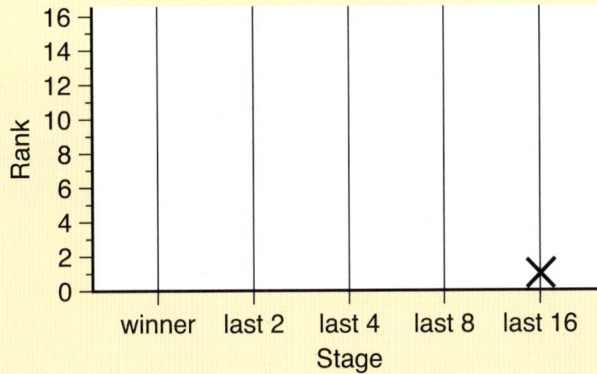

Team (and city)	Population	Rank	Stage in Champion's League
Arsenal (**London**)	7,600,000	1	Last 16
Barcelona	1,583,000		Last 16
Bayern **Munich**	1,332,000		Last 8
Celtic (**Glasgow**)	1,162,000		Last 16
Chelsea (**London**)	7,600,000		Last 4
Inter **Milan**	4,280,000		Last 16
Lille	1,730,000		Last 16
Liverpool	1,362,000		Last 2
Lyon	1,648,000		Last 16
Manchester United	2,586,000		Last 4
AC **Milan**	4,280,000		Winner
Porto	1,300,000		Last 16
PSV **Eindhoven**	750,000		Last 8
Real **Madrid**	3,228,000		Last 16
Roma (**Rome**)	2,648,000		Last 8
Valencia	796,000		Last 8

M European Champion's League, 2006–07

any questions...

Think of some geographical questions that you could ask now you have studied these pages. Think of a *WHERE?*, a *WHAT?*, a *WHY?* and a *HOW?* question. For example: *Why does a small city, like Eindhoven, have such a successful team?*

75

→ Field of dreams

Nowhere in the world is football growing faster than in Africa. Suddenly, African footballers are making a name for themselves across Europe. Their destination of choice is the English Premier League, which offers the greatest financial rewards. It is hard to imagine the journey that some African players make to reach the top of their profession.

Kolo Toure was born in Cote d'Ivoire, a country in west Africa (photo **N**). He did not own a pair of football boots until he was 15, often playing barefooted. Then he was spotted and joined the successful football academy at ASEC Abidjan, a club in the Cote d'Ivoire capital. It is known as the 'football factory'.

N Kolo Toure, Arsenal footballer

The journey from Cote d'Ivoire to Europe

O

● Arsenal paid just £150,000 for Kolo Tou a small price for a footballer these days. It is often cheaper for clubs to buy Africa players than European players.

ENGLAND
Arsenal

EUROPE

BELGIUM

● Once playing in Europe, the players are spotted by bigger European clubs and can be sold.

Thousands of African children dream of playing football in Europe

● The most gifted players are sent to Belgium to play for Premier League team, Beveren. The deal is that ASEC sends at least four players a year.

AFRICA

● ASEC Abidjan is one of the most success football academies in the world. As well as Kolo Toure, it has produced stars like Emmanuel Eboue, Salomon Kalou, Didier Sakora and Yaya Toure (Kolo's brother).

Abidjan

COTE D'IVOIRE

Not all dreams come true. For every Kolo Toure there are hundreds of African footballers you will never hear about. For some, the quest to become a professional footballer turns into a nightmare.

P An African footballer's story

Kenneth Akpueze, Nigerian midfield player, 21

'I flew from Nigeria to India. I was en route to Australia. An agent in Nigeria had said he would take me there to play; a good club, good money, everything. But in India, I couldn't get an Australian visa.

So I called the agent and he said he had a friend in Cambodia and that I had to go there to meet him, that they played very good football and that maybe I could make decent money there. I didn't have any alternative. I bought my ticket to Cambodia and met the friend. But he didn't know anything about football. He didn't even know if they played football.

I stayed here three or four months before I got a trial with a club and now I am here with Phnom Penh Empire. I never heard from my agent in Nigeria again. I am paid $250 (£150) a month here, that hardly takes me half way through the month. But I cannot return home. Africans never go home empty-handed. I cannot go home without making any money and I have to do something to take care of my family.

If I go home now, my brother will ask, "How about my money?" It was my brother who paid for my ticket here.

My younger brother rang recently saying that my sister is very unwell and that I have to send money for her to go to a hospital. But I don't have any money.

So I'd like to move. Anywhere would be better than Cambodia. I am planning to go to Singapore to play. I'd love to go to Europe, but nobody sees you in Cambodia. No one sees Cambodian football on TV, no agents come here – so how would anyone see us?'

Source: The Times (2007)

activity...

1 Follow the journey of African footballers on map **O**. What would be the benefits for:
  a) the African footballers who go to Europe?
  b) the African club they leave?
  c) the European club they join?
  d) European football fans?
  e) But, what might be the problems for:
    • African football fans?
    • European footballers?

discuss...

2 a) Why do you think ASEC Abidjan is called the 'football factory'?
  b) Some people say that sending African footballers to Europe is a form of modern day slavery. What do you think?

aim high...

3 Read story **P**. Imagine that Kenneth's younger brother, Samuel, also dreams of becoming a professional footballer, and asks him for advice.
Write a letter to Samuel, giving him advice about what he should do, or should not do, if he wants to make his dream come true.

any questions...

Think of some geographical questions that you could ask now you have studied these pages. Think of a *WHERE?*, a *WHAT?*, a *WHY?* and a *HOW?* question. For example:
Why do some countries, like Cote d'Ivoire, produce the best footballers?

→ Extra time!

The match – sorry, the unit! – is over. How many questions about football did you think of? Now you have to choose one of your questions and persuade your teacher that it is such a good geographical question that you should be given extra time to investigate it.

your final task...

1 Through the unit you have made a list of football questions. Now, you have to choose one question you can investigate with the help of geography. Do you remember the three rules to help you think of a good geographical question? They are going to help you now.

Rule 1 A question that geography will help you to answer. First, get rid of questions that have got nothing to do with geography, e.g. What is Wayne Rooney's latest car?

Rule 2 An open question that needs more than a one-word, or two-word, answer. Next, get rid of questions that need only a one, or two-word, answer, e.g. Where do Manchester United play?

Rule 3 A question that you REALLY want to know the answer to. Finally, look at the questions that you have got left. Choose the one that you are most interested in.

2 Once you've chosen your question, you have to persuade your teacher that it is a good geographical question. Again, the three rules can help you.

Take the example of Jordan's question, 'Why are there so many foreign players in England?' This is what he wrote to persuade his teacher that he should get extra time to investigate it.

My question is:
Why are there so many foreign players in England?

This is a geographical question because:
I will need to find out more about the countries that the players come from to find out why they want to leave. I will need to compare these countries with England to find out why they come here.

The question needs more than a one-word, or two-word, answer because:
It is a WHY? question. I need to find out why it happens. I don't expect to answer it in one or two words.

I really want to know the answer to this question because:
I support Chelsea and many of their players are foreign. I also want to be a footballer when I'm older. If there are lots of foreign players I might not get a chance.

Write your reasons to persuade your teacher that you have chosen a good geographical question. You can use the same writing frame as Jordan used to persuade his teacher.

6 Coast to coast
Where would you build a new seaside resort?

KEY CONCEP
- Space
- Environmental Interaction
- Physical proces

We live on an island with the sea all around us. It's not surprising lots of us like going to the seaside. And, with over 12,000 km of coastline in the UK, there are plenty of places to choose from.

coming up...

A property development company wants to build a new seaside holiday resort. You have been employed in a team of consultants to advise them where to build.

through the unit...

You will discover lots about coasts – the features you find there, the processes that change them, and some of the hazards – to help you to choose the best place. You will use Ordnance Survey maps to investigate four coastal areas.

your final task...

At the end of the unit you will make your final decision – where would you build your new seaside resort?

BEACHCOMBERS
Luxury property development

Dear Consultant,

Welcome to the team of consultants for our newest luxury coastal development.

We are looking for the ideal coastal site to build an exclusive new seaside holiday resort with five hundred homes, a hotel, shops and leisure facilities. We have identified a shortlist of four possible sites around the UK (photographs of the sites are attached to this letter). Your job, working with the team, is to decide on the best site and then write a report to explain your choice.

You will need to consider the following factors before you come to your decision:

<u>Landscape</u> The site should have good sea views and attractive coastal scenery nearby.

<u>Safety</u> It should be safe from possible hazards like cliff falls or flooding.

<u>Future coastline</u> It should not be affected by any future changes in the shape of the coastline.

<u>Environmental impact</u> It should not have a harmful effect on the environment.

I look forward to receiving your final report with your decision. Good luck!

Yours sincerely,

Project Director

①

Marloes, Pembrokeshire

Happisburgh, Norfolk

WHERE WOULD YOU BUILD A NEW SEASIDE RESORT?

Porlock Bay, Somerset

Winchelsea, East Sussex

starter...

1 Look at the four photos. What are your first impressions?

a) Which place would you most like to be in?

b) What do you like about it? (You will need more ideas than this before you make your decision. The company are not paying you lots of money just to say which place you like!)

activity...

2 Here are some activities that you can do at the seaside:

sunbathing   swimming   cliff-walking
surfing   bird watching

Which do you think would be the best place to do each of these activities? In each case, give reasons.

81

Mapping out the coast

Before you start your job, you and your
team visit the four coastal areas to get to
know more about them. Here are the maps
for your fact-finding trip.

A Porlock Bay, 1:25,000 OS map

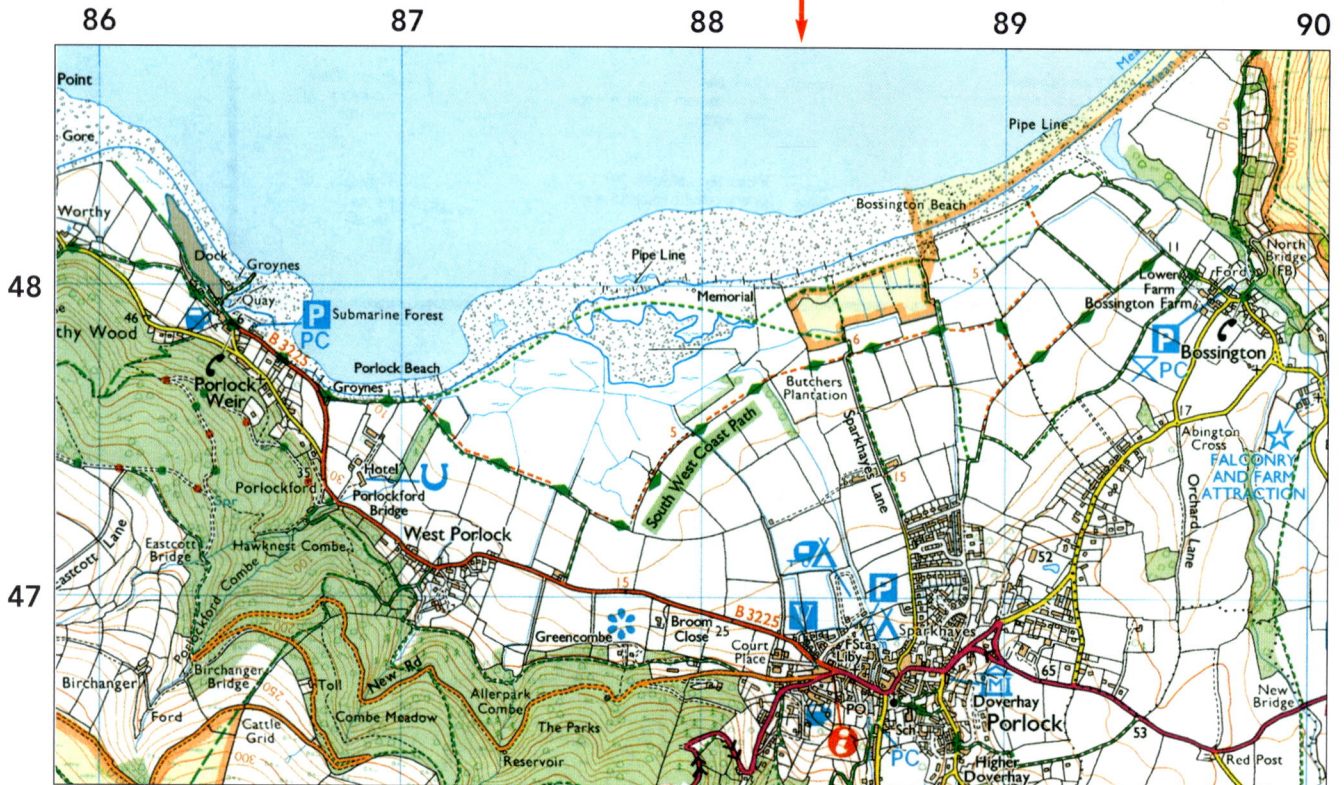

B Marloes, 1:25,000 OS map

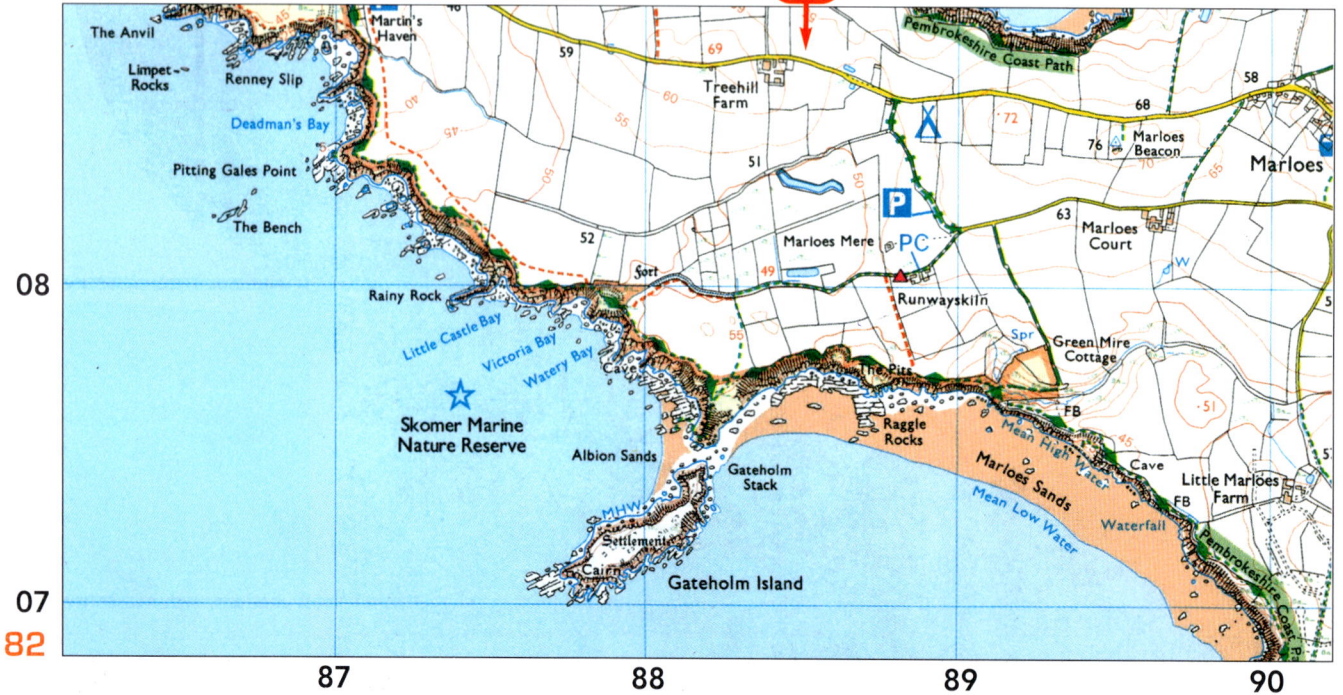

C Happisburgh, 1:25,000 OS map

D Winchelsea, 1:25,000 OS map

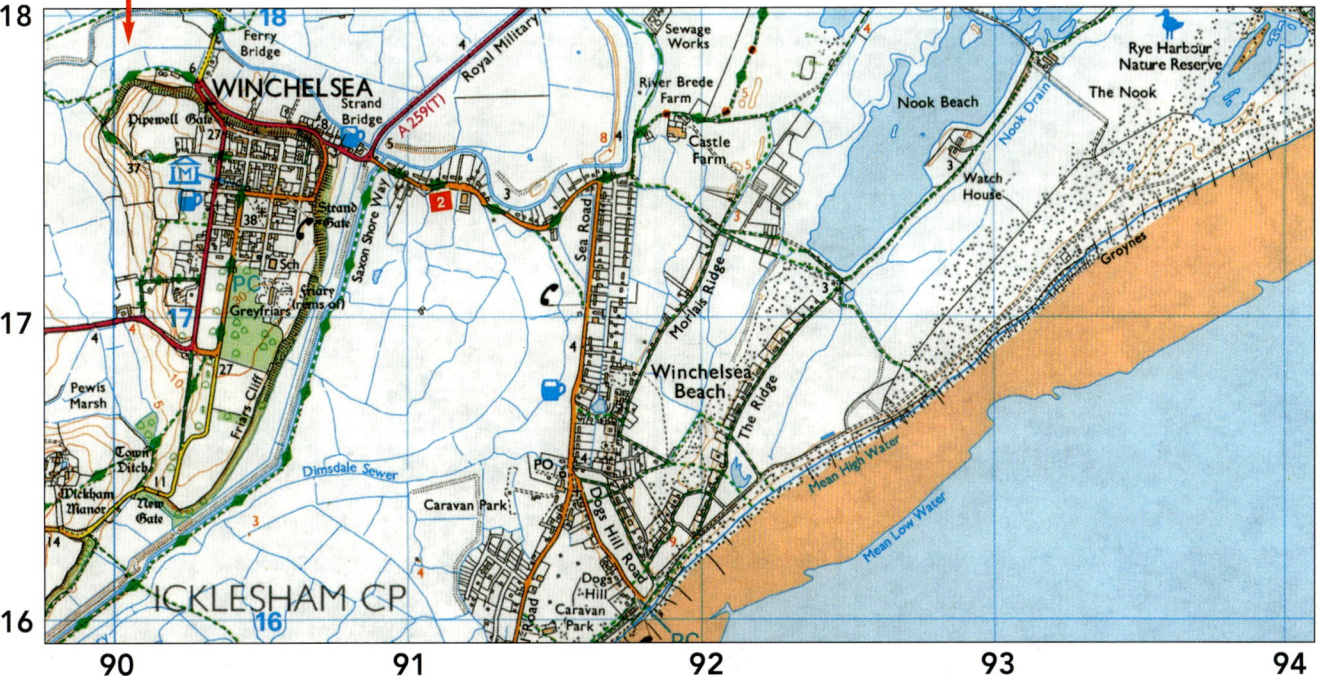

activity...

Read the following statements and then carefully examine the maps. For each statement:

- decide which place you are most likely to be in: Porlock Bay, Marloes, Happisburgh or Winchelsea?
- write a sentence to explain your answer.

a) Stay overnight in a Youth Hostel.

b) Climb a hill to a historic town.

c) Look out of the window to see a lighthouse.

d) Get pebbles stuck between your toes on the beach.

e) Go to sleep in a caravan to the sound of waves.

f) See a dolphin swimming in the sea.

g) Trip over an old tree stump on the beach.

h) Feel a cold north-easterly sea breeze.

→ A new seaside resort

The seaside is back in fashion. For years, traditional British seaside resorts have been declining. Now, people are coming back to the seaside, not just for holidays but to live. City-dwellers want a quiet place to relax at the weekend, active people like sailing and surfing in their spare time, and older people often retire by the sea.

So, what could a modern seaside resort look like? Here is a computer-generated image of what Porlock Bay might look like with a new resort.

E Porlock Bay as it looks now

F A computer-generated image of Porlock Bay, as it could look with a new resort

consultant's report...

1 Work in a group of four (you will work with them throughout this unit).

a) Each of you is a consultant. You each specialise in one of the four sites. Allocate one site to each consultant:

b) Draw a large table to cover two pages in your book, like this:

Porlock Bay Marloes
Happisburgh Winchelsea

Site	Landscape	Safety	Future coastline	Environmental impact
Porlock Bay				
Marloes				
Happisburgh				
Winchelsea				

As you go through the rest of this unit, make notes in your table about your site.

(At the end of the unit, you will share your notes with your team to complete the table. Then you will decide on the best site and write your report.)

c) Look at the photo of your site on page 80 or 81. Describe the landscape in the photo, and fill in column 2 of your table.

activity...

2 Look at photo **E** and image **F**.

a) Either: Draw an outline sketch of photo **E**. Then add an outline of the new resort in image **F**. Or Download photo **E** onto a computer screen. Then add image **F** onto photo **E** on the screen.

b) Do you think this is a good site for the resort? Look at your picture and give a reason.

aim high...

3 You are going to choose another site for a seaside resort.

a) Choose one of the other photos on pages 80 to 81. (You could choose the site for which you are consultant.)

b) Draw a sketch of the photo (or download the photo onto your computer screen).

c) Choose a suitable site for a new resort. Add an outline of the resort onto the sketch or photo.

d) Do you think this would be a good site for a resort? Give a reason.

→ Shaping the coastline

The coastline is shaped by the action of waves on the land.
This is how they do it:
- they ERODE, or wear away the land.
- they TRANSPORT the eroded material.
- they DEPOSIT, or drop, the material somewhere else.

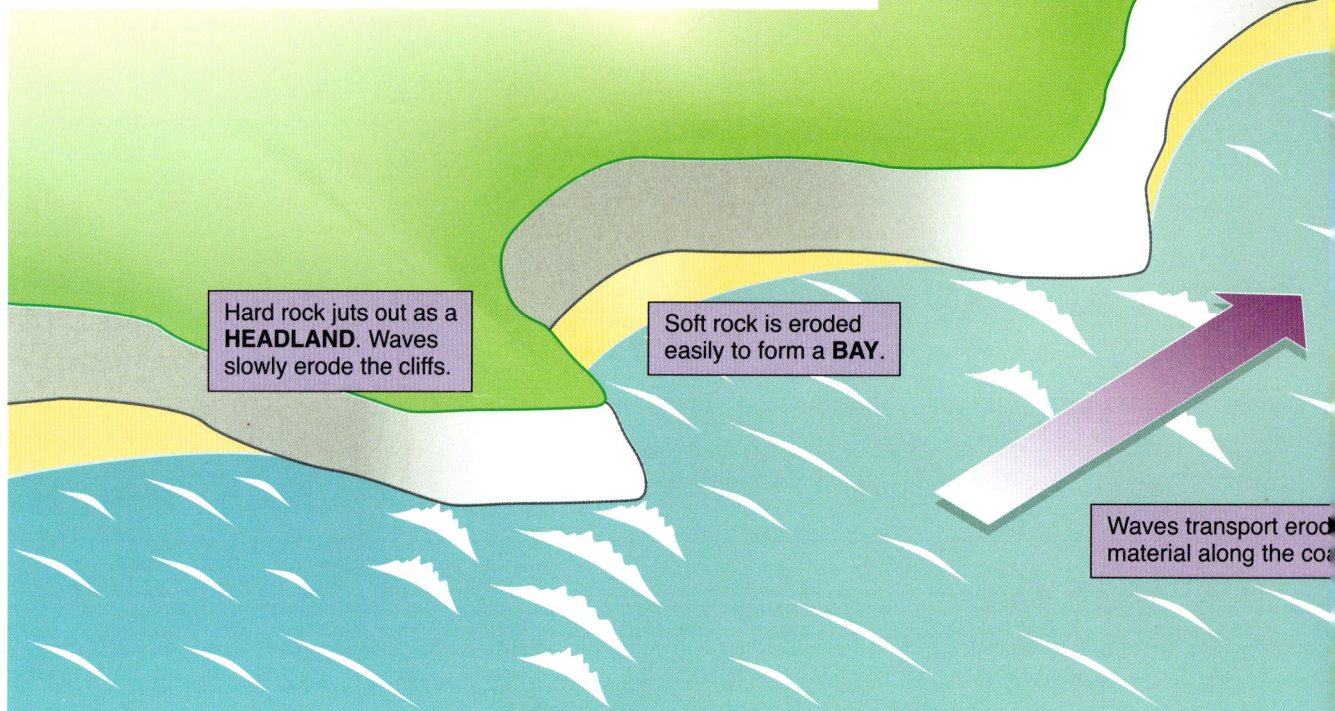

Hard rock juts out as a **HEADLAND**. Waves slowly erode the cliffs.

Soft rock is eroded easily to form a **BAY**.

Waves transport erod
material along the coa

activity...

1 Look at map **G**.
Find these grid references on the map:
885 483   878 478   875 488
For each one, would you be on a headland, in a bay, on a beach, or in a marsh?

2 Compare map **G** with map **H**.
  a) What type of rock is Hurlstone Point made of?
  b) What type of rock lies under the village of Porlock?
  c) Which type of rock is eroded most easily: sandstone or clay? Explain how you can tell from the coastline.

3 Choose a site on map **G** to build a new seaside resort.
  a) Cut out a piece of paper this size and shape. (Keep this to use on other maps later.)
  b) Move it about on the map until you find the best site. The site should be:
  - close to the sea with a good view
  - fairly flat (contour lines are far apart)
  - safe from flooding (at least 5 metres above sea level).
  Give a 4-figure grid reference for the site you choose.
  c) Explain your choice.

Another headland – and more erosion. On and on around the coast...

Waves lose their energy. They deposit sand and shingle to form a **BEACH**.

Some coastlines run straight for miles and miles, while others wiggle in and out. Why? It's all to do with the geology, or types of rock. Where the coast has a mixture of hard and soft rocks, bays and headlands are likely to form. Map **G** shows Porlock Bay in Somerset. Compare it with a geological map of the same area (map **H**).

G 1:25,000 Ordnance Survey map extract, Porlock Bay

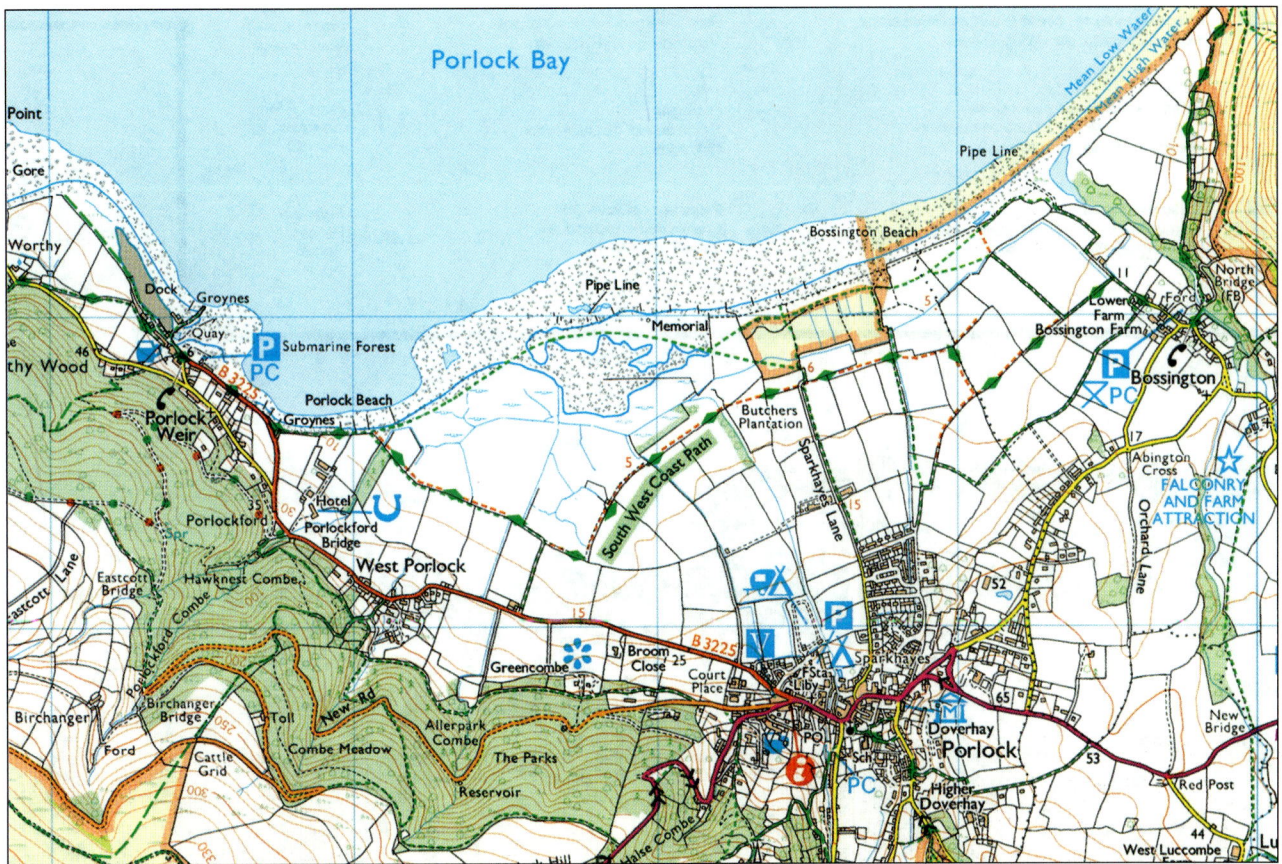

H Geological map of Porlock Bay

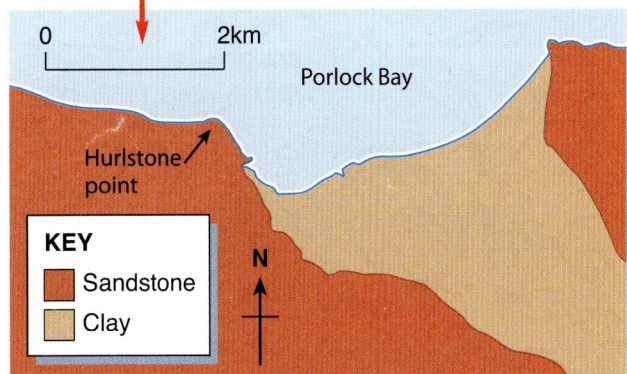

KEY
Sandstone
Clay

consultant's report...

4 What coastal landscape features can you find on the map? List them in column 2 of your table.

(If you are the consultant for Porlock Bay use map **G**. Maps for the other three sites are on pages 82 to 83.)

→ Wind and waves

Waves are created by wind blowing over the surface of the water. Try blowing over a bowl of water to prove it for yourself.

It's the same with the sea. Wind blows over the sea and ruffles the surface to make waves. The size of the waves depends on:
• the *strength* of the wind
• the *length* of time the wind blows for
• the *distance* over which the wind blows – called the FETCH.

Around the British Isles the biggest waves come across the Atlantic Ocean, which has the longest fetch (map **I**). The bigger the waves the more energy they have to shape the land. And this is how they do it …

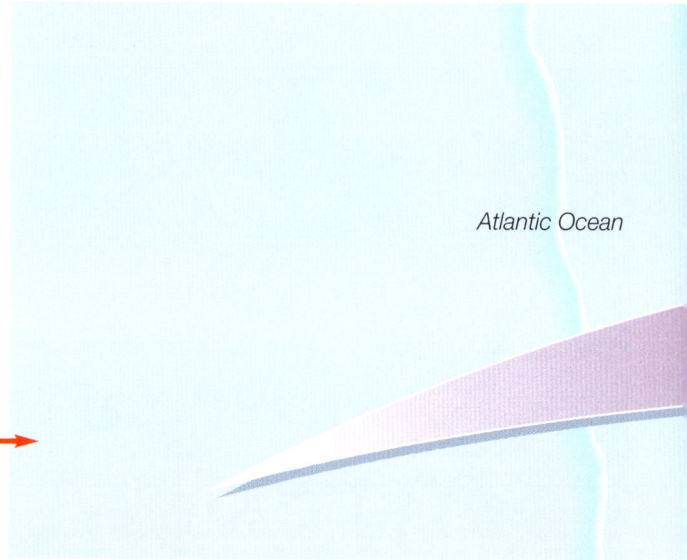

I →

Winds around the British Isles

Atlantic Ocean

Hydraulic action

I'm cracking up under the pressure!

Corrasion

If you can't beat them join them, I say!

Attrition

All this erosion is wearing me down!

Solution

Where am I?

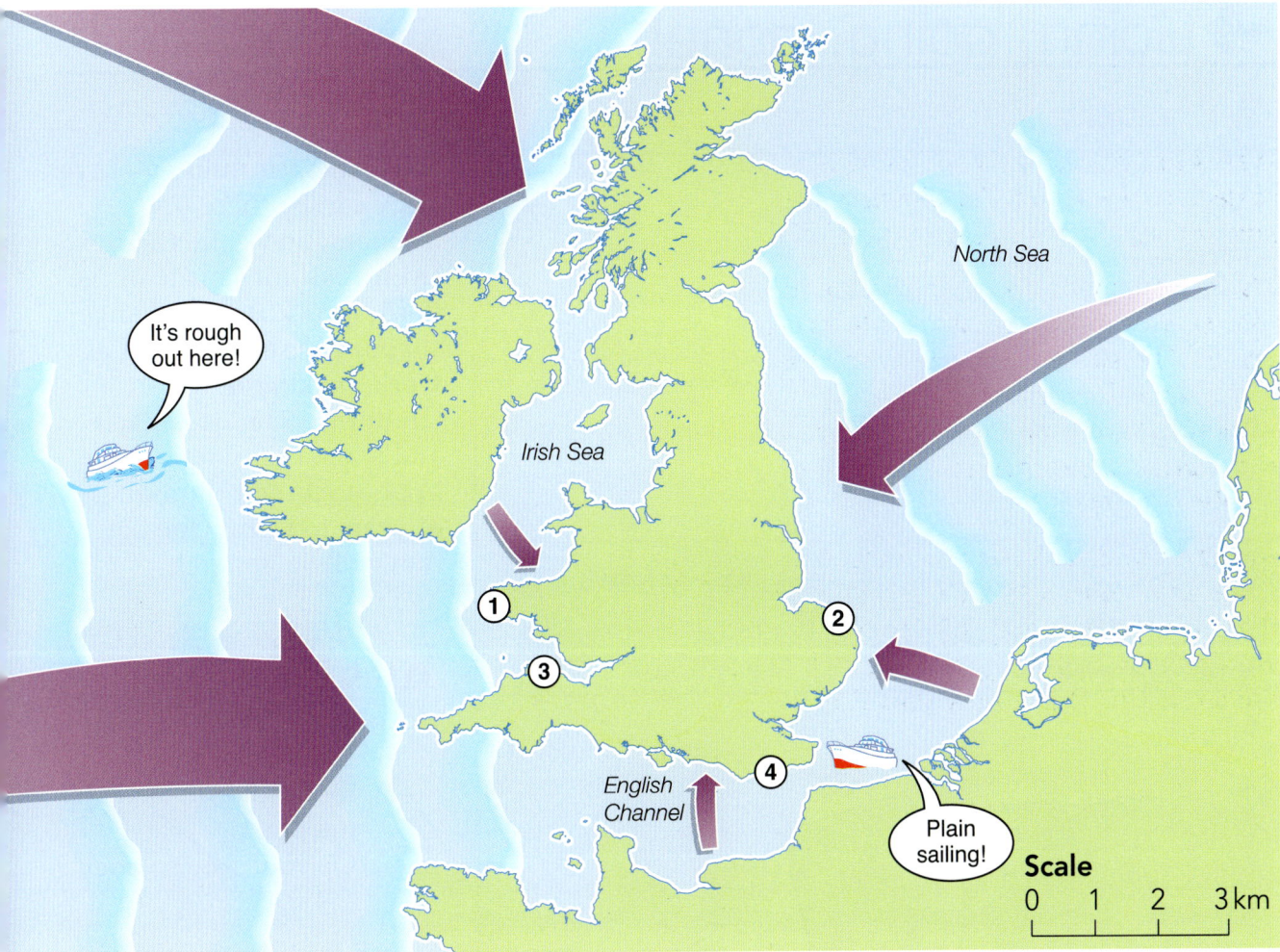

North Sea

It's rough out here!

Irish Sea

① ② ③ ④

English Channel

Plain sailing!

Scale
0   1   2   3 km

activity...

1 Look at the drawings of wave erosion. Match each drawing with one of these descriptions. Write them in your book with the correct heading.

  a) Rocks and stones bump into each other and get worn down to sand or shingle.

  b) Waves hammer into cracks in the rock, forcing it to split apart.

  c) Some rocks dissolve slowly in seawater.

  d) Waves throw rocks and stones at the cliff, wearing it away like sandpaper.

2 Look at map **I**. Find the four possible sites for your seaside resort on the map. Which site (1, 2, 3 or 4) would you expect to have: **a)** the biggest waves **b)** the smallest waves? In each case, explain why.

3 Look again at the photos of the four sites on pages 80 to 81. Imagine your site on a windy day. The waves are bigger, and the sea covers more of the beach.

  a) Draw a sketch to show what you think it would look like now.

  b) Would it be a safe place to build a seaside resort, do you think? Explain why.

consultant's report...

4 How windy would your site be? How safe would it be to build the resort here? Write you ideas in column 3 of the table.

➡ Going, going, gone!

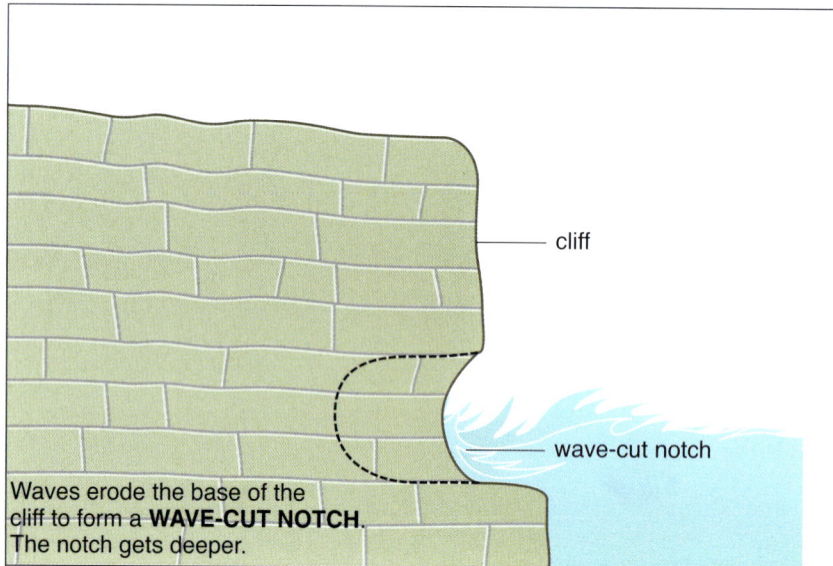

cliff

wave-cut notch

Waves erode the base of the cliff to form a **WAVE-CUT NOTCH**. The notch gets deeper.

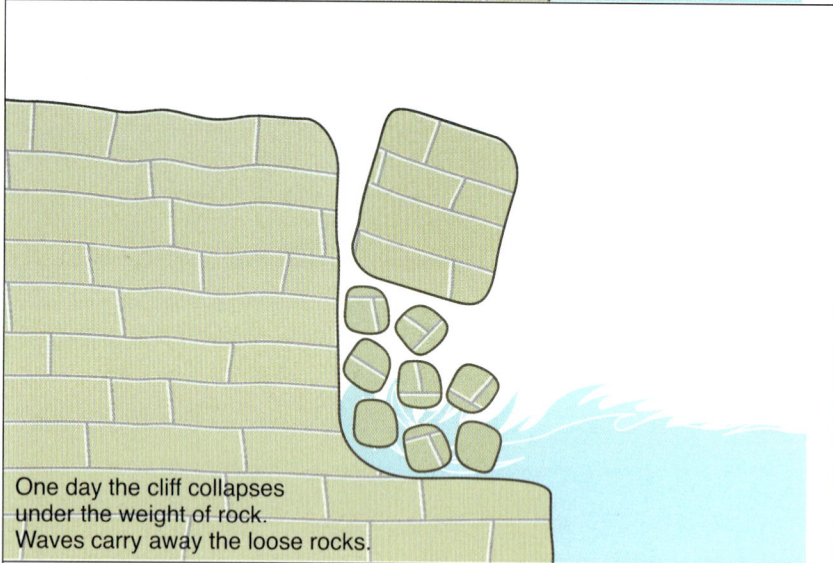

One day the cliff collapses under the weight of rock. Waves carry away the loose rocks.

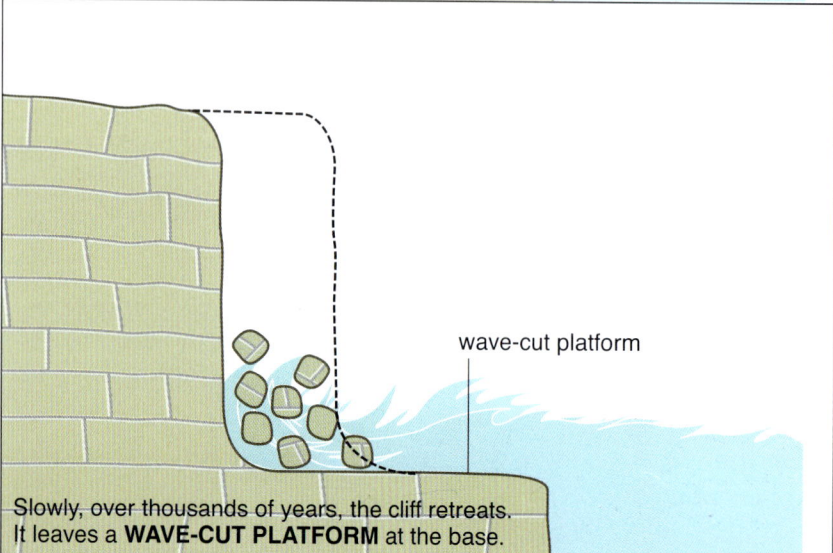

wave-cut platform

Slowly, over thousands of years, the cliff retreats. It leaves a **WAVE-CUT PLATFORM** at the base.

Don't mess around on CLIFFS. They can be there one moment, and gone the next. Even coastlines made of the hardest rock slowly change shape. The sea is always at work, making new landforms. The CAVE, ARCH and STACK that you can see in photo **J** took thousands of years to form. One day they will be gone.

activity a...

1 Look at photo **J**.
  a) Draw a sketch of the landforms in the photo.
  b) On your sketch, label:

cliff   cave   arch

stack

cliff

cave

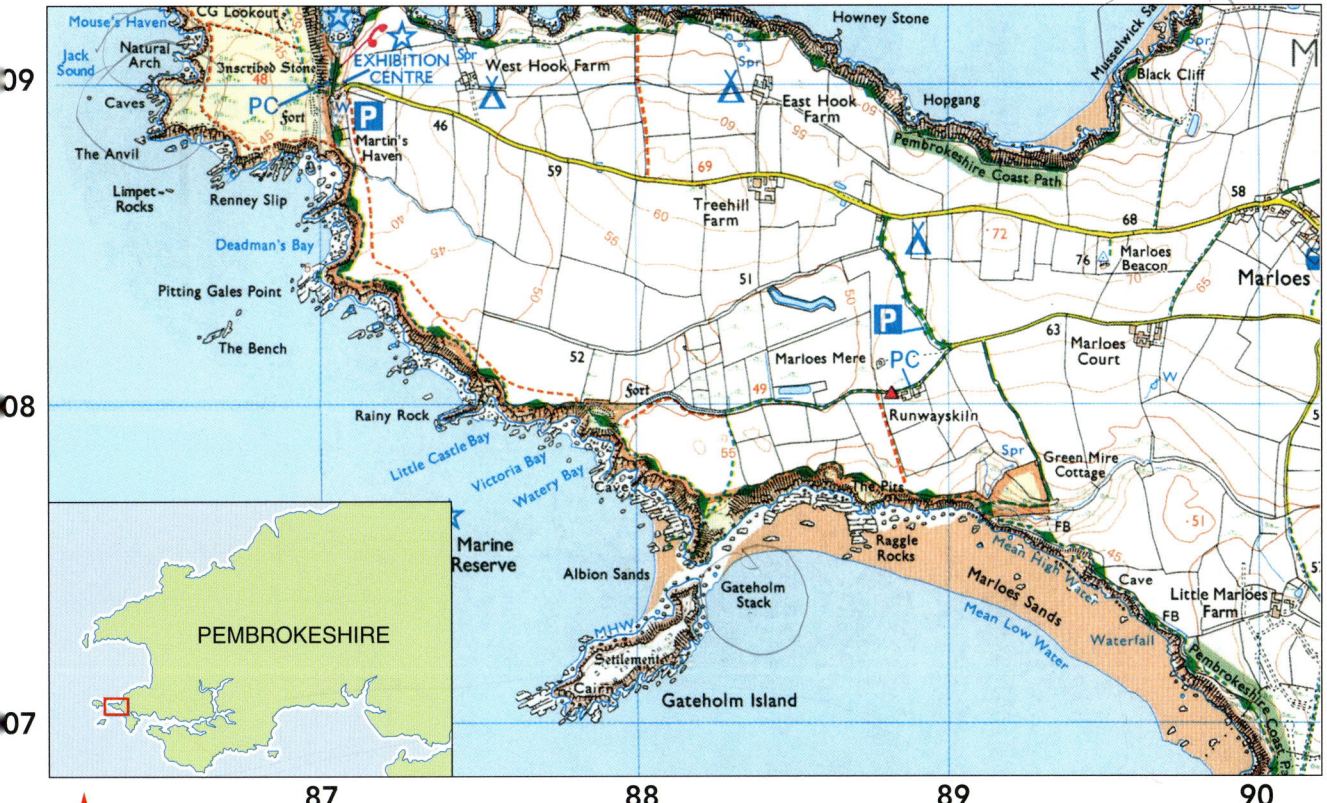

K 1:25,000 Ordnance Survey map extract, Marloes, Pembrokeshire

J Coastal scenery in Pembrokeshire

activity b...

2 Read the sentences below. They explain how the landforms in photo **J** were formed. Re-write the sentences in the correct order.
  a) The roof of the arch collapses.
  b) Waves erode the base of the cliff at the headland.
  c) The cave wears through the headland, forming an arch.
  d) A wave-cut notch is formed.
  e) A stack is left standing on its own.
  f) The notch becomes deeper to form a cave.

3 Look at map **K**.
  a) Find an example of each landform (a cliff, a cave, an arch and a stack). Give a 6-figure grid reference for each one, e.g. *cave at 770 077*
  b) Choose a site on the map to build a seaside resort. You can move your piece of paper around the map to help you to decide. Give a 4-figure grid reference.
  c) Explain your choice.

consultant's report...

4 Is the site near a cliff? How safe would it be to build a seaside resort here? Write your ideas in column 3 of your table. (If you are the consultant for Marloes use map **K**. Maps for the other three sites are on pages 82 to 83.)

91

→ Shifting sands

Some changes at the coast take forever – others happen as you sit and watch. Look at the pebble on a beach in drawing **L**. It is moving in and out with the waves. If you have time to sit all day, you will see the pebble move along the beach.

This process is called LONGSHORE DRIFT. And it's not just one pebble that is moving – the whole beach is on the move! The sea transports pebbles, shingle and sand along the coast day after day, year after year.

How longshore drift works

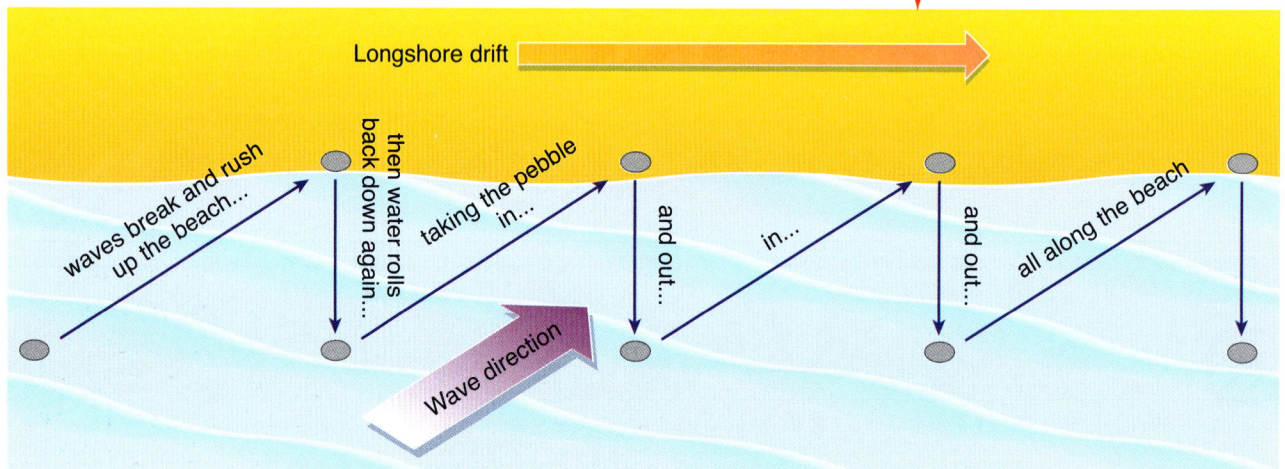

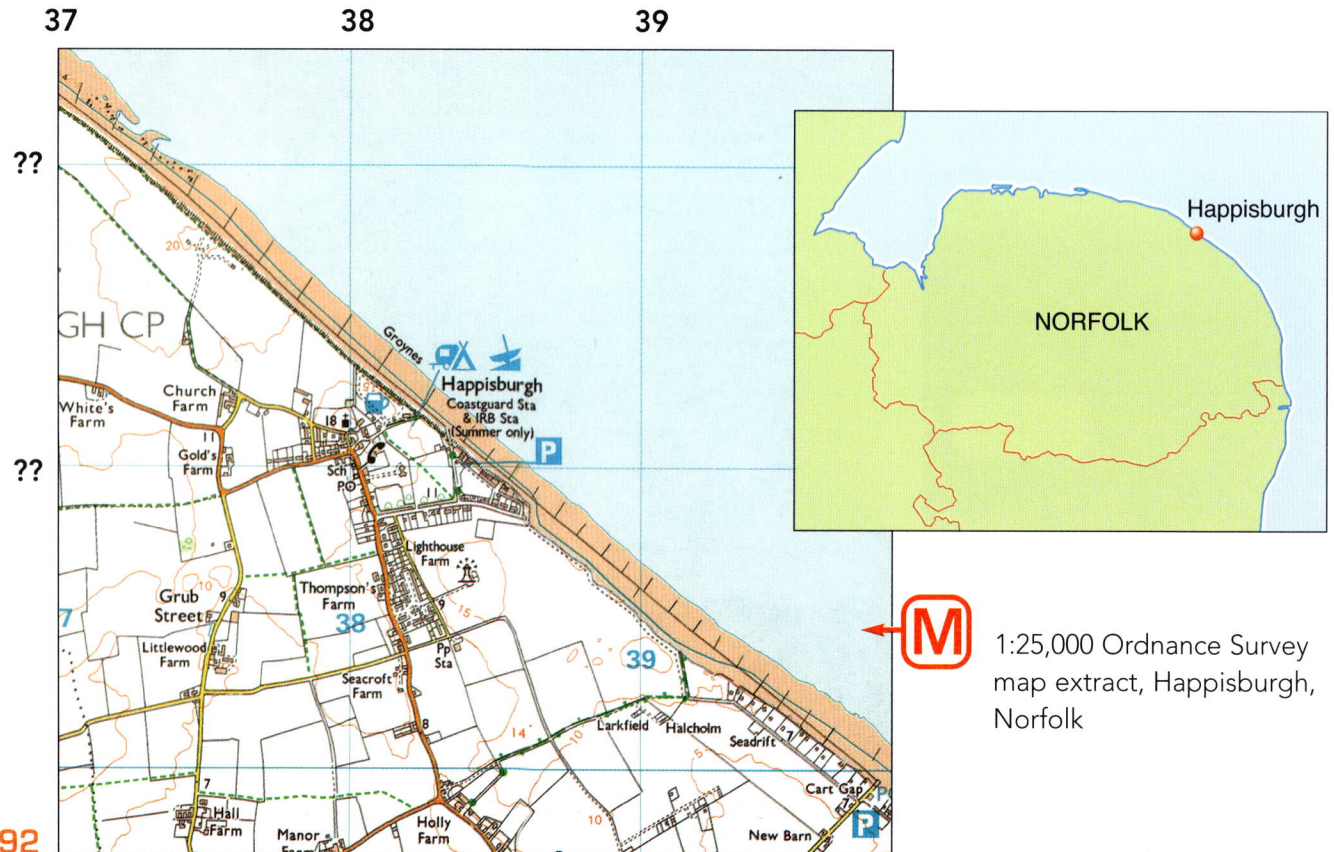

Longshore drift

waves break and rush up the beach...

then water rolls back down again...

taking the pebble in...

and out...

in...

and out...

all along the beach

Wave direction

1:25,000 Ordnance Survey map extract, Happisburgh, Norfolk

Happisburgh

NORFOLK

In seaside towns in Norfolk the council have built GROYNES to prevent longshore drift (photo **N**). The idea is to stop the waves from taking away the beach. But, it doesn't completely stop longshore drift …

… further down the coast in Suffolk is Orford Ness (photo **O**). This is a long tongue of land called a SPIT. The sea transports the material across the mouth of a river and deposits it here. The river is diverted southwards.

N Groynes on a beach in Norfolk

prevailing wind blows in this direction

sand builds up behind the groyne

N

groyne

sea transports material along the coast and deposits it here

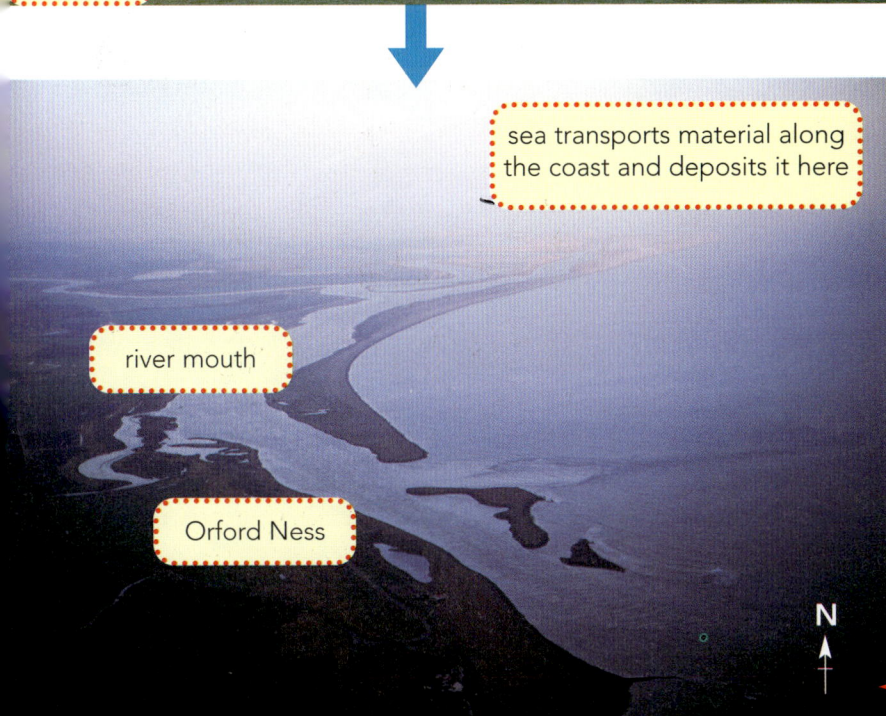

river mouth

Orford Ness

N

O Orford Ness in Suffolk

1 Look at photos **N** and **O**.
  a) What evidence of longshore drift can you see in:
    i) photo **N**? and
    ii) photo **O**?
  b) In which direction is longshore drift moving sand along the coast? Explain how you can tell from the photos.
  c) Explain the connection between the photos.

2 Look at map **M**.
  a) Choose a site on the map to build a new seaside resort. Try to avoid knocking down any buildings. You can move your piece of paper around the map to help you to decide. Give a 4-figure grid reference.
  b) Explain your choice.

aim high...

3 Compare the coastline on map **M** with the coastline on map **K** on page 91. The photos on pages 80 to 81 will also help.
  a) Describe at least three differences between the coastlines.
  b) Can you suggest reasons for these differences?

93

➡ On the brink of disaster

Some parts of the coast are eroding more quickly than others. Over the centuries many villages on the east coast have been lost to the sea. A few more metres of land disappear every year. The village of Happisburgh (pronounced Hays-borough), in Norfolk, could be the next to go.

P Article from the *Eastern Daily Press*

23rd December 2003

COASTAL EROSION LEAVES HOUSE ON THE EDGE

By Maria Fulcher

A teashop and guesthouse in Happisburgh is under threat after the coast took another battering in weekend storms.

Di Wrightson has been running the business with Jill Morris for 23 years, but the storms have left them wondering how much longer their livelihood will survive. When Miss Wrightson first moved into the house on Beach Road there were properties between her and the sea, but now only a metre separates her rear garage from the waves below.

She said, 'We shall continue trading until we are forced to go and we are hoping to open in the New Year if we are still here.'

A combination of enormous waves and heavy rain led to the latest LANDSLIDE, taking another chunk of cliff into the sea. The coastline at Happisburgh is made of soft sand and clay. When the rock becomes saturated with water, the cliffs collapse.

Miss Wrightson said, 'We shall miss it terribly. We are going to lose everything we have worked for and will have to go and rent somewhere. We certainly cannot afford to go and buy a house.' The government offers no compensation to people who lose their homes due to coastal erosion, and no insurance company will cover such a property. Around 15 homes and 7 caravans have already been lost to the sea along the Happisburgh coastline in the last 11 years.

Q Happisburgh

Why can't something be done to save Happisburgh? It can, but the problem is the cost. Box **R** shows some of the methods that can be used to protect the coast.

R Choices for coastal protection

Build a **SEA WALL**. It is the best way to stop erosion, but also the most expensive. Cost: £5,000 per metre.

Build wooden **REVETEMENTS**. Like a sea wall, they break the force of the waves. They are made of wood and won't last so long. Cost: £2,000 per metre.

Build groynes. They slow down longshore drift and keep the beach in place. This protects cliffs from erosion. Cost: £10,000 each (200 metres apart).

Make a **ROCK BARRIER**. Piles of huge rocks protect the base of the cliffs. Cost: £1,000 per metre.

Regularly undertake **BEACH NOURISHMENT**. Replace sand and shingle lost from the beach. Cost: £25,000 per kilometre each year.

Do nothing and watch the sea erode the coastline. Costs nothing!

activity...

1 Read newspaper article **P**.
  a) Identify Di and Jill's house in the photo. Imagine that you live here. How would you feel?
  b) Write a letter to the government from Di appealing for more help for Happisburgh. What do you think should be done?

2 Update what has happened at Happisburgh since 2003. Check the website of the Coastal Concern Action Group at www.happisburgh.org.uk. Has Di Wrightson's house gone? What else has disappeared? What protection has been built?

aim high...

3 a) Work out how much it would cost to protect 1 km of coastline for 50 years, using each method in box **R**. Complete a table like this.

Method	Cost	Length/number/years	Total cost
Sea wall	£5,000 per metre	1000 metres	£5,000,000

  b) Which method should be used to protect Happisburgh? Give reasons.

consultant's report...

4 How quickly is the coastline eroding? Does it need protection, and how much would this cost? Write your ideas in column 4 of your table.
  If you are the consultant for Happisburgh use map **M** on page 92. Maps for the other three sites are on pages 82 to 83.

→ Now you see it ...

Throughout history, our island has been changing shape. Most changes have been due to the action of the sea. Some changes have been due to people.

Romney Marsh, in the south-east corner of England, was once a shallow bay. The village of Winchelsea was a port on the coast. Since the Middle Ages, people have RECLAIMED Romney Marsh from the sea. They built a sea wall around the bay and drained the marsh behind it. Winchelsea is now high and dry, two kilometres inland. Romney Marsh has become an important area for farming and wildlife. But for how much longer?

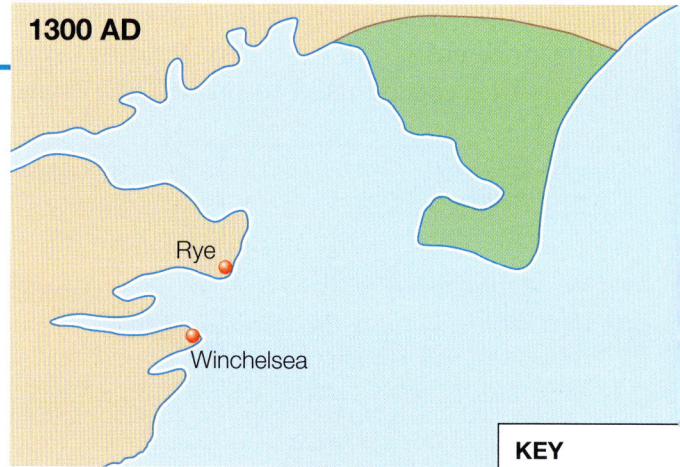

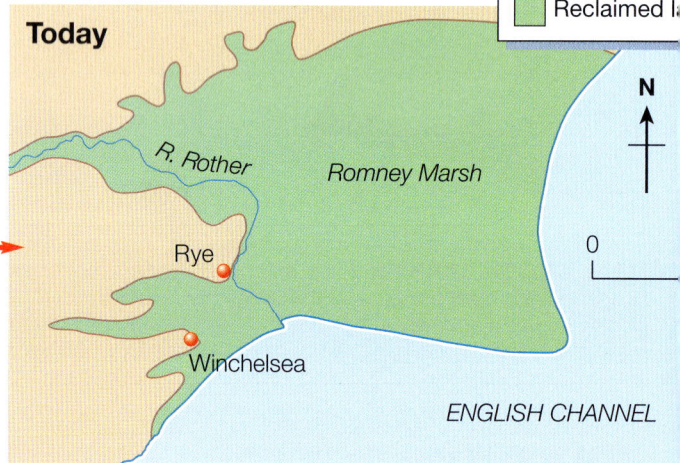

1300 AD

Rye

Winchelsea

KEY

Land over 5

Reclaimed l

N

Today

R. Rother

Romney Marsh

Rye

Winchelsea

ENGLISH CHANNEL

0

S →
Reclamation of Romney Marsh

T
1:25,000 Ordnance Survey map extract, Winchelsea, East Sussex

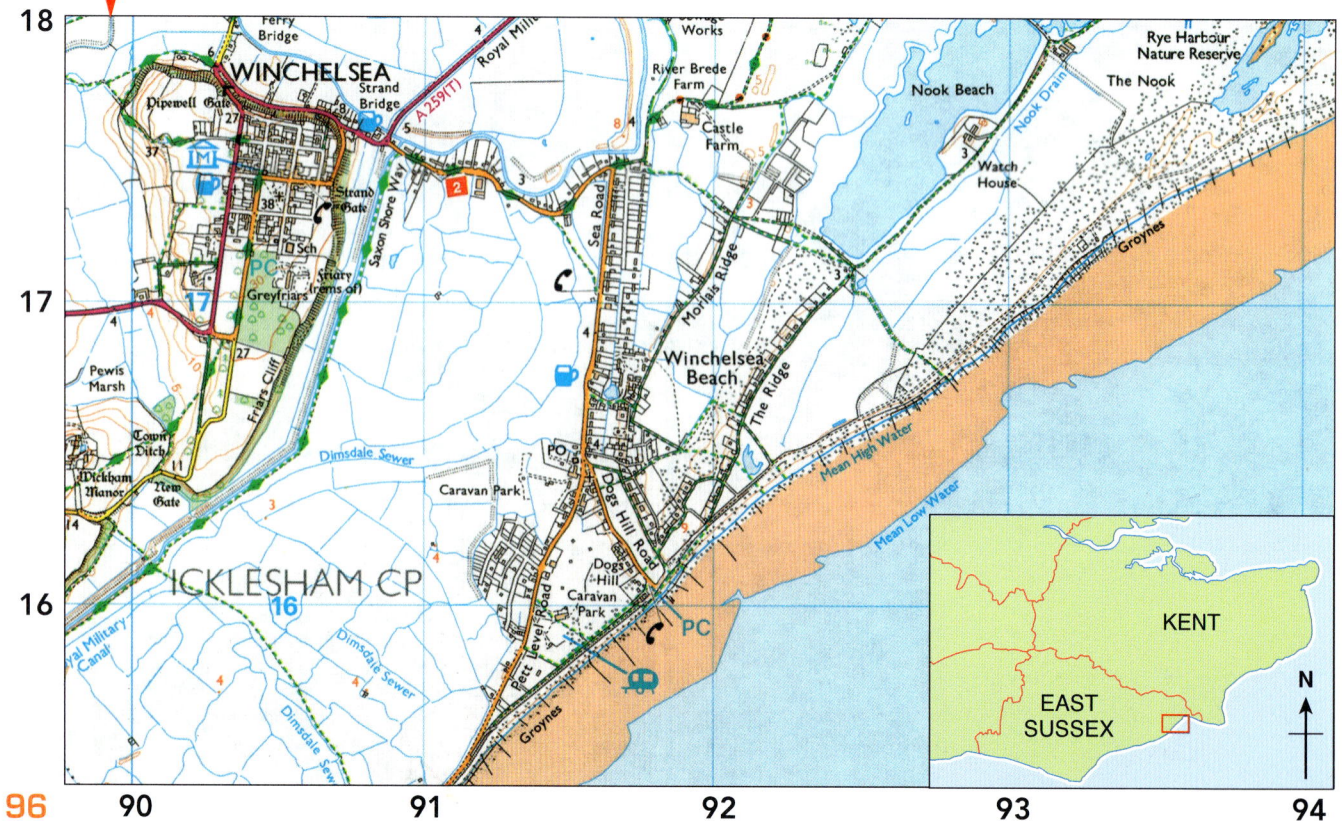

KENT

EAST SUSSEX

N

→ ... Now you don't

Our battle to keep out the sea could soon get harder. Sea levels around the world are rising with global warming (you can find out more in Unit 8). Scientists predict that it could rise by up to four metres over the next 100 years. If that happens, the areas of Britain shaded on map **U** are in danger of flooding.

We will have to decide which areas are worth protecting. To protect Romney Marsh a new sea wall would have to be built, and that could be very expensive (look back at page 95).

Flood danger **U** in Britain

KEY

Areas at risk from flooding

Dornoch Firth

Firth of Clyde

Solway Firth

Morecambe Bay

The Fens

Norfolk Broads

Severn Lowlands

Somerset Levels

The Solent

Romney Marsh

activity...

1 Look at map **T**.
  a) Draw a sketch map, like the one below, showing the modern coastline. Colour the sea blue.

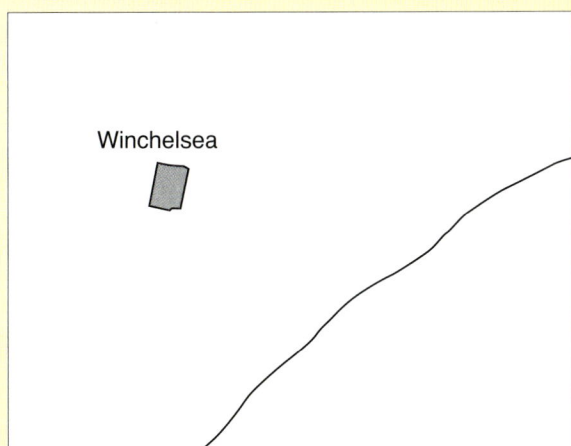

Winchelsea

  b) Find the old coastline around Winchelsea. A rough guide is the 5 metre contour line, labelled Friars Cliff.
  c) Now, add the old coastline to your map. Shade the land in two colours. Shade the land over 5 metres brown. Shade the reclaimed area, below 5 metres, green. Give the map a key.

2 Look again at map **T**.
  a) Identify at least five buildings or human activities that would be lost if Romney Marsh was flooded. Give grid references. For example, *Castle Farm at 919 176.*
  b) Should a new sea wall be built to protect Romney Marsh, or should it be allowed to flood when sea levels rise? Explain your opinion.

3 a) Choose a site to build a new seaside resort on map **T**. You can move your piece of paper around the map to help you to decide. Give 4-figure grid references.
  b) Explain your choice.

consultant's report...

4 How much of the area is below 5 metres? Is it in an area that is likely to flood in future? (Map **U** will help you.) Write your ideas in column 4 of your table. (If you are the consultant for Winchelsea use map **T**. Maps for the other three sites are on pages 82 to 83.)

→ Impact on the environment

This is the computer-generated image you saw earlier, showing what a new seaside resort could look like. What impact would it have on the environment? While many people like to go to busy seaside resorts, others go to the coast to enjoy beautiful natural scenery. Building a new seaside resort could spoil the view. It could also create more traffic and harm wildlife living there. In some places there are laws to protect the environment by preventing any new building. National Trust land and nature reserves are protected in this way.

Computer-generated image of the new seaside resort

activity...

1 Look at image **V**. It shows the impact of a new resort at Porlock Bay.
   a) Do you think the new resort makes the environment better or worse? Explain why.
   b) How could you reduce any harmful impact it has?
2 Look at the maps on pages 82 to 83.
   a) What protected areas can you find on the maps? Look for symbols for National Trust land and nature reserves. Give 4-figure grid references for each one.

   b) Is the site that you have chosen to build a seaside resort in a protected area? If so, choose a new site on the map. Give a 4-figure grid reference for it.

consultant's report...

3 Imagine the impact of a new resort on the environment around your site. Would a new resort make the environment better or worse? How could you reduce any harmful impact? Write your ideas in column 5 of your table.

■ your final task...

With your team of consultants, you are going to make your final decision about the best site to build a new seaside resort. Then you will report back to the property development company.

1 Share your notes with the other members of your team to complete your table.

Either Take turns to read your notes for the team to copy them into the table.

Or Type your notes into a copy of the table on a computer.

Factor \ Site	1	2	3	4
Landscape				
Safety				
Future coastline				
Environmental impact				
TOTAL				

2 Together with your team, complete a score chart (like the one here) to compare the four sites. You need to agree on the scores before you write them down. The top scores go to the site that:
- has the best landscape
- is in the safest location
- is least likely to be affected by rising sea level
- has least impact on the environment.

Give a score, from 1 to 4, to the four sites for each factor. For example, if **you think that** site 1 has the best landscape **give it a score of** 4 **in the top left box in the score chart.**

3 Add together the scores for each site to work out a total score. Write the total scores in the bottom row of the score chart.

Which site gets the top score? This is the site that you will choose to build the new seaside resort.

4 Write a report for the property development company to explain your choice.
- Why did your chosen site come top? Mention the factors where your chosen site scores well.
- Why did you not choose the other three sites? Mention the factors where the other sites score badly. You can use this writing frame to help you.

Report on the best site for a new seaside resort

Consultant team . . . .

Date . . . .

The site chosen by the team to build the new

seaside resort is . . . .

. . . .

The factors in favour of this site are . . . .

. . . .

. . . .

. . . .

The sites rejected were . . . .

. . . .

. . . .

We did not choose . . . .

. . . .

because . . . .

. . . .

. . . .

7 Spain – here we come!

What won't you see on a package holiday to Spain?

coming up...

Have you ever been to Spain? How much did you find out about the country? You'll be travelling around Spain to meet people and find out more about the country.

through the unit...

As you travel, you will start to plan your next holiday. You might even like Spain so much, you decide to stay!

your final task...

You will use your knowledge of Spain to plan your next holiday, but this time you might see more of the country.

Fiestas – they know how to party in Spain

Many British people think that Spain is just sun, sea and sand. But, there's a lot more to Spain than that. Here are some of the things that you would miss if you went on a package holiday to the beach …

Modern cities

Dangerous sports!

Eating out – Spanish people do it a lot

Snowboarding in the mountains

Spain

Real Madrid – some say the best football team in the world

starter...

1 a) Think about Spain. What images come into your mind? Make a list of at least five. (Your teacher probably told you to do this before you opened the book.)

b) Now, look at the photos on these pages. How do they differ from your images? Did any of them surprise you? Why?

c) Looking at these photos, would you want to go to Spain for a holiday? Why, or why not?

2 What do you know about Spain? For example: *people go there on holiday, the capital is Madrid.* Draw a spider diagram to show what you know.
Organise your ideas into groups, like this:

Where is it? *in Europe*
Similarities with the UK *big cities*
Place *Madrid*
Differences from the UK *speak Spanish*
SPAIN
Climate *hot summers*
Also famous for *Real Madrid*
Famous people *Picasso, the artist*
Tourist attractions *the coast*

As you go through the unit you will find out more about Spain. You can add more ideas to your diagram.

101

→ Spain – our favourite country

Spain is the most popular holiday destination for people in the UK (graph **A**). Each year 13 million of us visit Spain. There are also nearly a million British people who own a property in Spain. Many of them live there all year round.

Meanwhile, most Spanish people prefer to stay in Spain for their holiday. Many city dwellers have a second home in the country or near the coast. So what is it about Spain that everybody – both British and Spanish – seem to like?

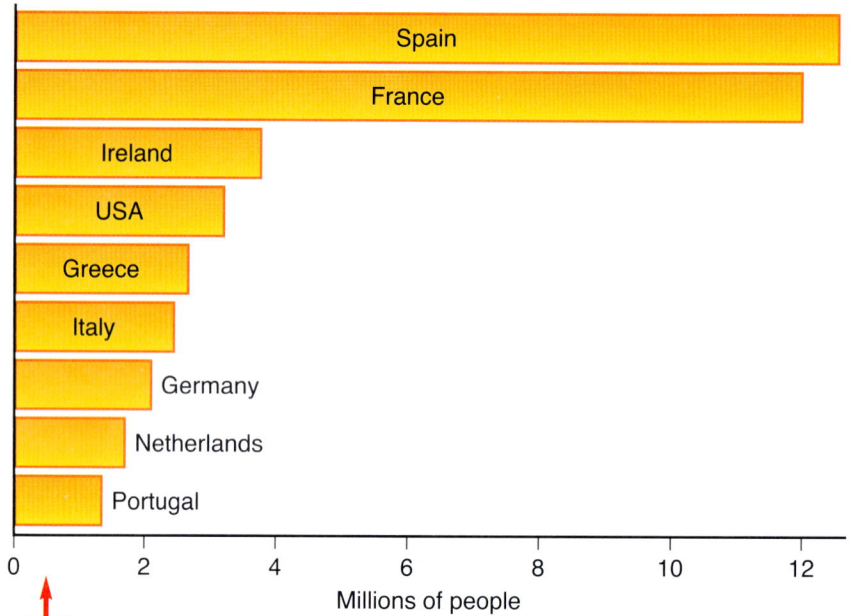

A Most popular holiday destinations for UK tourists

Source: Office for National Statistics

activity...

1 Study box **B**. Then, carry out a survey of pupils in your class.
   a) Find out what things would most attract them to Spain.
   b) Draw a bar chart to show the results of your survey. Draw a bar for each of the things that attract them.
   c) What is it that would most attract pupils in your class to Spain?
2 Box **B** shows the things that British and Spanish people like about Spain (but, they don't necessarily like the same things).
   a) Draw a large diagram like this and label the three areas.

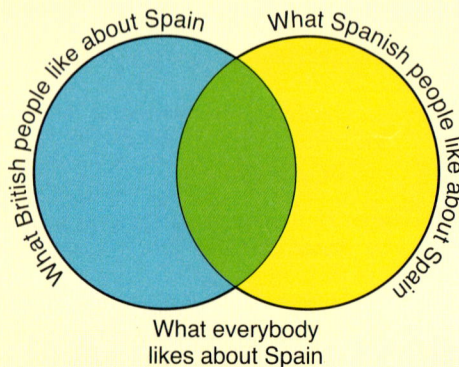

What British people like about Spain

What Spanish people like about Spain

What everybody likes about Spain

   b) Think about what British and Spanish people might like about Spain. Sort the things into three groups. List them under the headings in your diagram.

What *do* I most like about living in Spain? It's the streetlife. People don't stay at home. We go out to meet each other in bars, cafes or in the plaza (town square). I never get lonely in Spain.

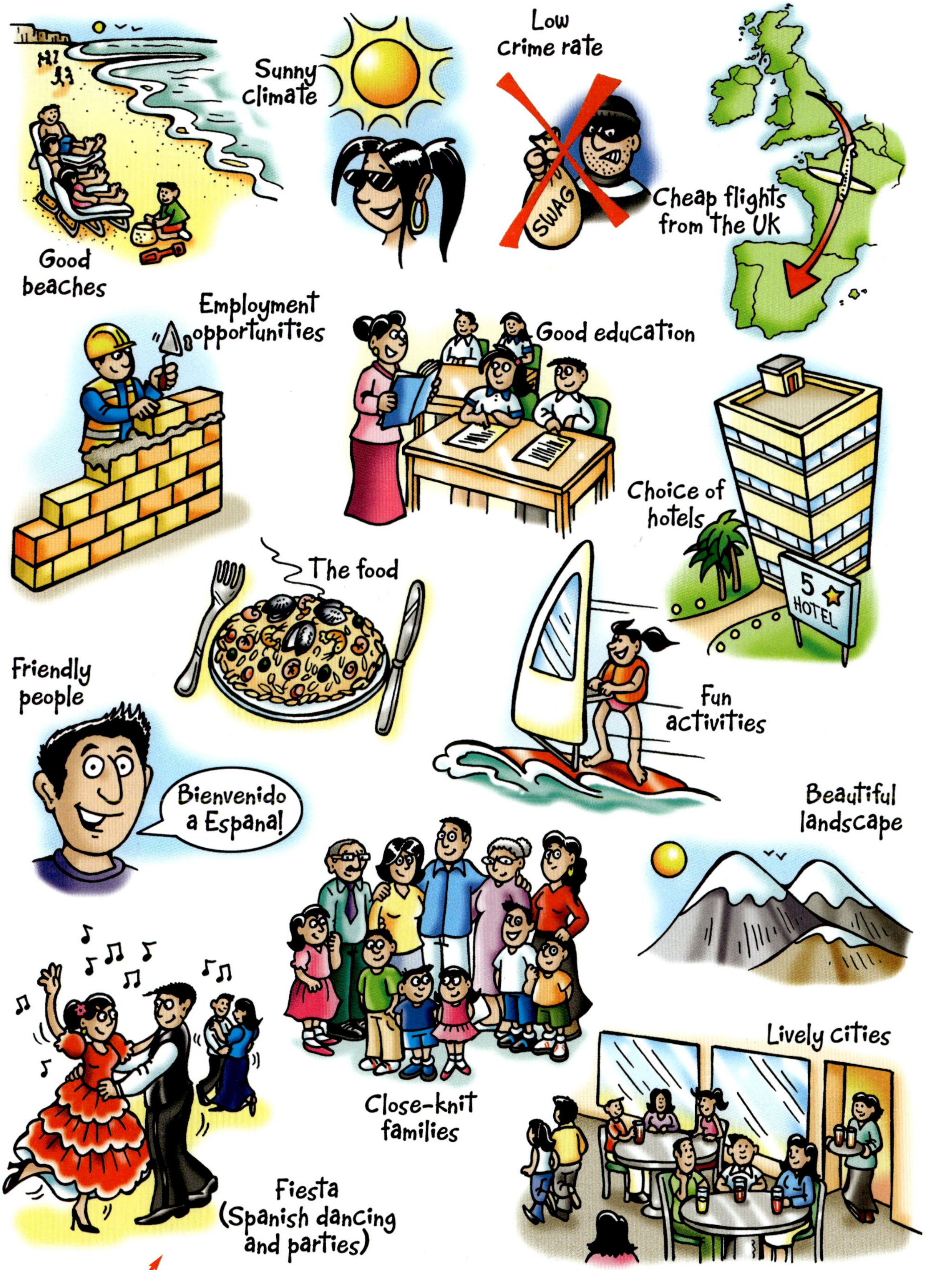

Good beaches
Sunny climate
Low crime rate
SWAG
Cheap flights from the UK
Employment opportunities
Good education
Choice of hotels
5 HOTEL
The food
Friendly people
Bienvenido a Espana!
Fun activities
Beautiful landscape
Close-knit families
Lively cities
Fiesta (Spanish dancing and parties)

B Things people like about Spain

➡ The Spanish jigsaw

Take human features, like countries, cities and towns away from a map, and what have you got left? The answer is a physical map. Physical maps show natural features like mountains and hills, rivers and lakes.

Here are six physical REGIONS of Spain (map **C**). They are muddled up and the wrong way round.

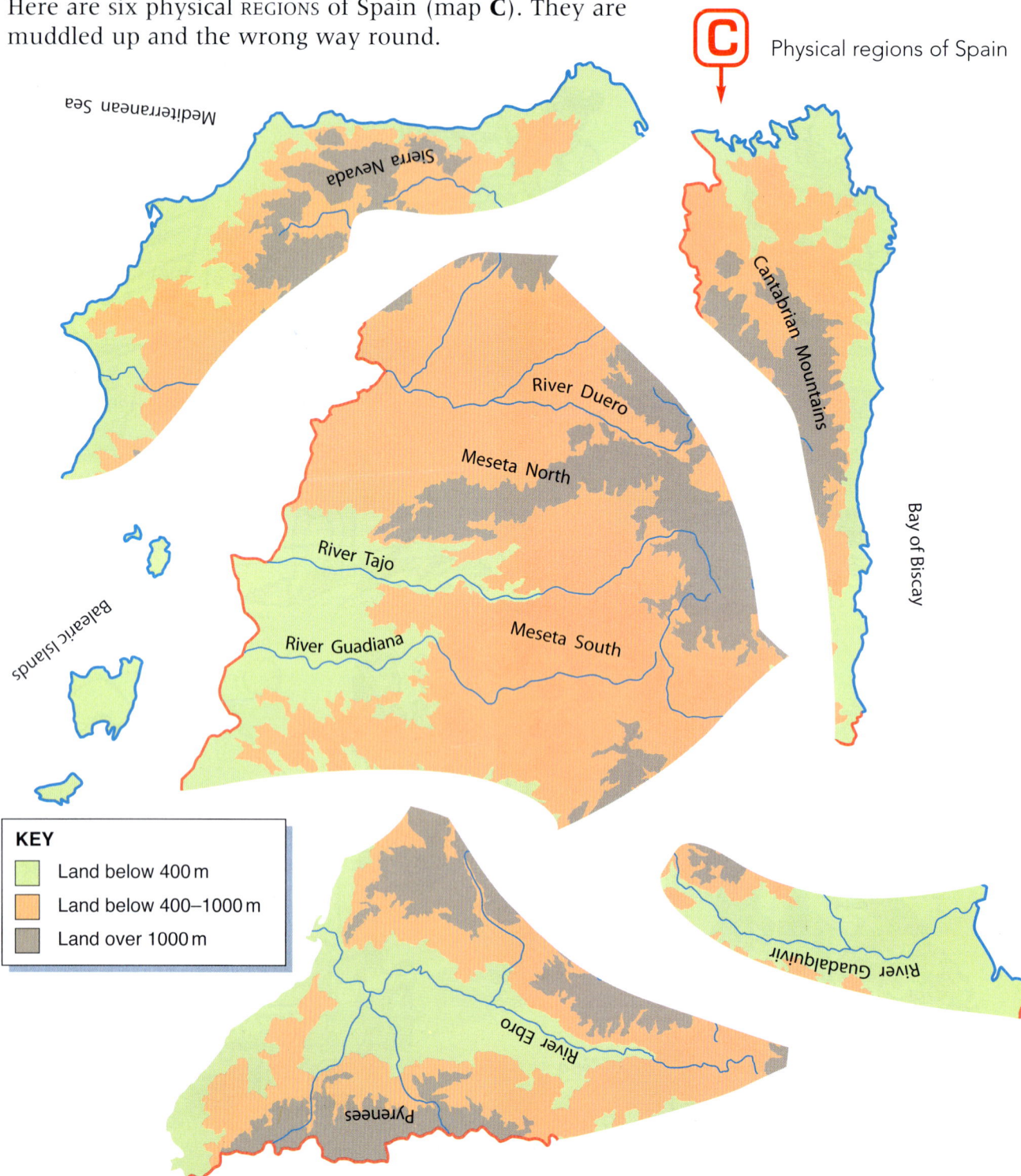

C Physical regions of Spain

Mediterranean Sea

Sierra Nevada

Cantabrian Mountains

Bay of Biscay

River Duero

Meseta North

River Tajo

Balearic Islands

River Guadiana

Meseta South

KEY

	Land below 400 m
	Land below 400–1000 m
	Land over 1000 m

River Guadalquivir

River Ebro

Pyrenees

Cantabrian Mountains,
Northern Spain

Meseta (the vast plateau that
covers central Spain)

Guadalquivir Valley

Mediterranean Coast

Balaeric Islands

Pyrenees

activity...

1 Look at the jigsaw of the physical map of Spain.
   a) Trace and cut out copies of the jigsaw pieces in **C**.
   b) Put the pieces together on a page in your book to complete a map of Spain.
   c) Stick the pieces down and give your map a title.

2 From what you see in the photos, in which regions of Spain would you be likely to do each of these activities:

★ climbing   ★ sailing
★ cycling   ★ skiing
★ swimming   ★ surfing
★ walking ★ bird-watching

holiday plans...

Find out more about one activity that you would like do in Spain. Go to the Spanish tourism website at www.spain.info

- choose an activity to do in Spain
- write it in the 'Search' bar and click 'Search' (a list of links should appear)
- click on the links to find a place in Spain where you could do your activity.

Some visitors think that Spain is one long beach. But, did you know that Spain is the most moutainous country in Europe (after Switzerland)? We've also got rivers, islands and even desert!

105

→ Sunny Spain?

'The rain in Spain falls mainly on the plain', or so the rhyme goes. But is it true? More to the point, when you plan your holiday, can you believe the brochure when it says 'Sunny Spain'? Let's find out.

D Types of climate in Spain

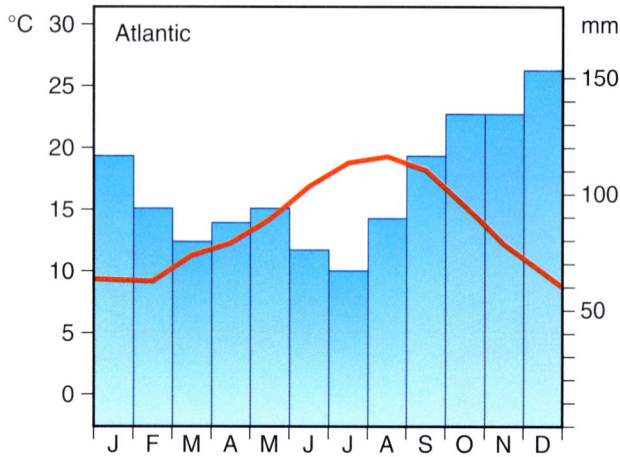

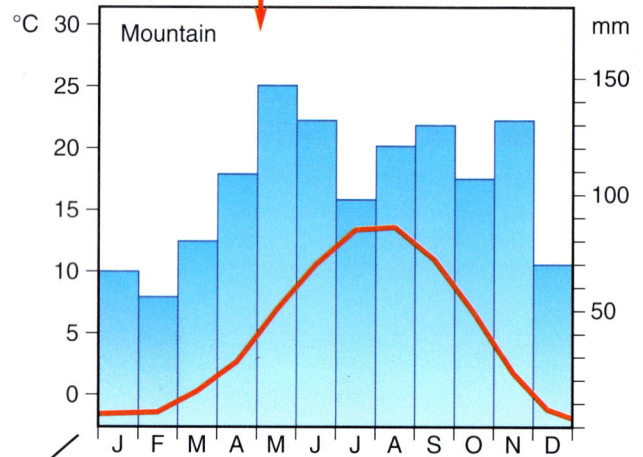

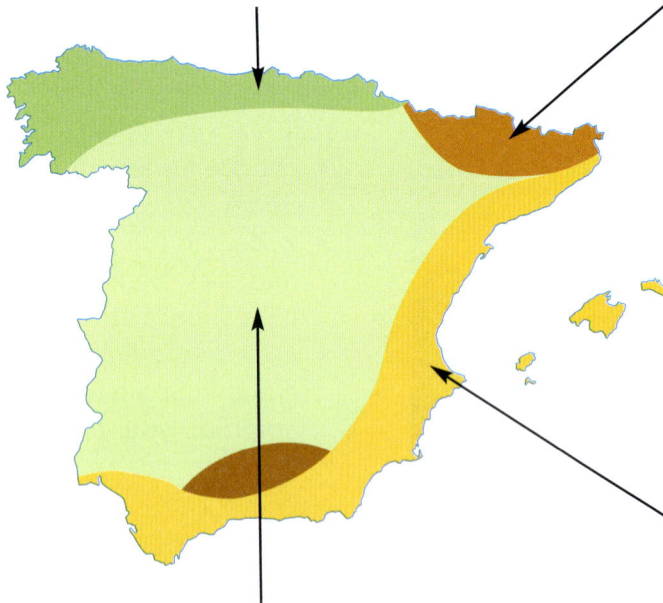

The graphs show the climate in different parts of Spain. Climate is the average weather pattern through the year. If you study the graphs carefully you can work out what the weather is likely to be on your holiday (no guarantees!). The blue bars represent rainfall in mm, the red line temperature in °C.

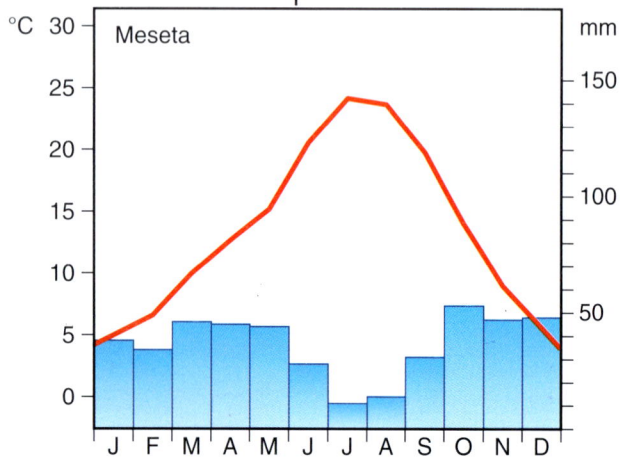

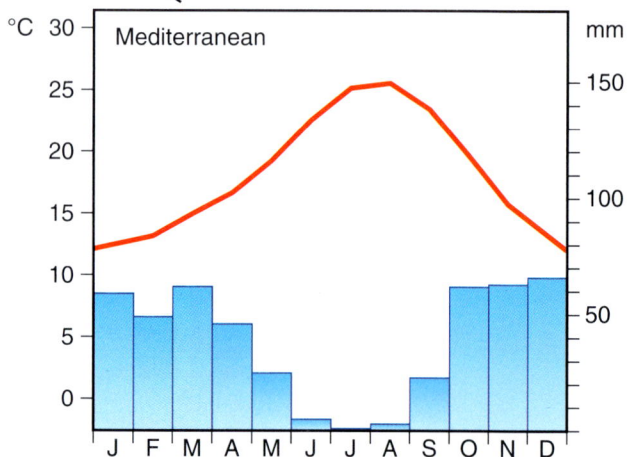

Key
- Mountain climate
- Meseta climate
- Atlantic climate
- Mediterranean climate

E The Meseta region of Spain

aim high...

3 Study the graphs around map **D** carefully.

a) Choose the correct phrases from the box to describe the four types of climate in Spain. Choose up to five phrases for each climate.

> ★ hot summers ★ warm summers ★ cool summers ★ freezing winters ★ cold winters ★ mild winters ★ high annual rainfall ★ low annual rainfall ★ wet winters ★ dry winters ★ wet summers

b) Describe the four types of climate in Spain.

holiday plans...

Think about the activity that you chose on page 105. Find out what the climate is like in the place where you will do it. Decide what would be the best time of year to do the activity. (For example, if you went cycling in the Meseta it would be too hot in summer and too cold in winter, so you could go in spring or autumn.)

activity...

1 Look at the climate graphs around map **D**. Read these statements about climate in Spain. Are they true or false? Use the information in the graphs to help you.
- The north of Spain is drier than the south.
- Farmers in the centre of Spain need to water their crops in summer.
- The Mediterranean coast is cooler than the Atlantic coast.
- The mountains are where it is most likely to snow in winter.

Copy the statements that you think are true. Re-write the statements you think are false, to make them true.

2 Look at photo **E**.
a) What does the photo tell you about the climate in the Meseta?
b) 'The rain in Spain falls mainly on the plain.' True or false? (Clue: Plain is another word for the Meseta.) The climate graph on map **D** will help you.

> The sun doesn't always shine in Spain. Where I come from, in the north-west of Spain, the weather is more like Britain. It rains all year round. If you want sunshine you need to go to the Mediterranean.

→ City break

Spain is almost exactly twice the size of the UK, yet it has fewer people. Large areas of the country are almost empty, but you'd never guess if you were in the middle of Madrid. Which city will you go to for your holiday, or will you try to get away from it all?

F A map of Spain showing both relief (height of the land) and population

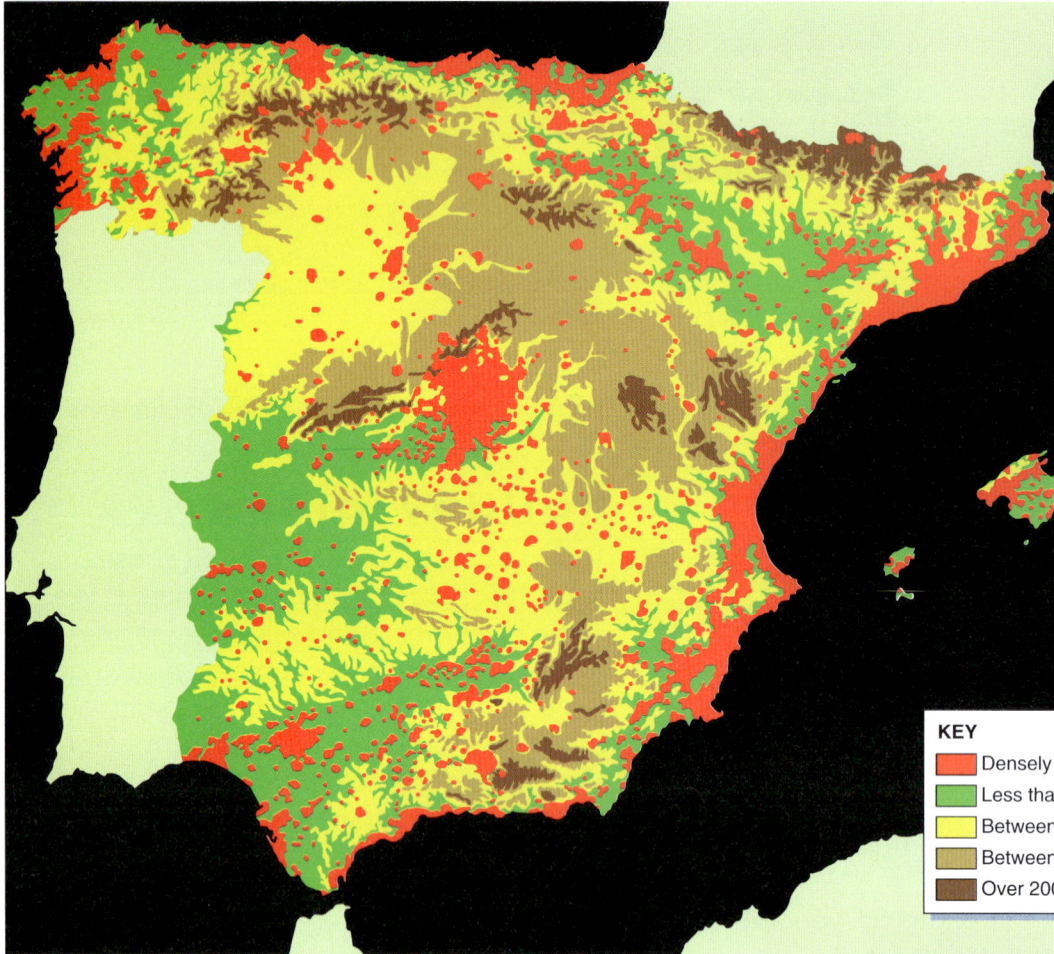

KEY

🟥	Densely populated
🟩	Less than 500 metres
🟨	Between 500 and 1000 metres
🟫	Between 1000 and 2000 metres
🟤	Over 2000 metres

City	Population
Madrid	3,228,000
Barcelona	1,583,000
Valencia	796,000
Seville	695,000
Zaragoza	601,000
Malaga	542,000
Bilbao	351,000
Murcia	350,000
Valladolid	317,000
Cordoba	307,000

Our cities are a bit different to yours. Spanish cities go to sleep in the afternoon and come alive at night! Shops and businesses close down for the afternoon when it can get really hot. But, in the evening, everything opens up again and people come out.

G Spain's top ten cities

There's lots to do in cities …

Visit the sights. The Alhambra palace in the southern city of Grenada.

activity...

1 Compare Spain and the UK.

	Spain	UK
Population (millions)	44	61
Area (square kilometres)	506,000	245,000

a) Calculate the population density in the two countries.
(Remember: population density = population/area)

b) Which country is the most crowded?

2 Look at the cities in table **G**.

a) Can you find these cities on satellite image **F**? (Cities lie at the centre of the areas with high population density.) Check the locations, using a map of Spain on page 135.

b) Mark and label the cities on an outline map of Spain. You can add more cities. Try to locate them as accurately as possible.

aim high...

3 Describe the population distribution in Spain, using map **F**. Mention each of these areas in your description:

> ★ Mediterranean coast ★ Meseta
> ★ Atlantic coast ★ mountain regions
> ★ river valleys (e.g. the Ebro valley)

Go shopping. The famous Ramblas shopping street in Barcelona.

holiday plans...

Which city would you most like to visit in Spain? The photos will give you some ideas. You can get more ideas from this website www.spain.info

a) Choose a city to visit.

b) Write it in the 'Search' bar and click 'Search'.
A list of links for the city should appear.

c) Click on the links to find out more about the city.

Watch a football match. Real Madrid play at the Bernabeu Stadium.

➡ Made in Spain

(H) Spain's regions

You may have already noticed, there isn't really one Spain – there are lots of Spains! Each region that you can see on the political map (**H**) has its own culture and identity, just as regions like Wales do in the UK. Some regions in Spain even speak their own language – others are identified by what they produce. La Rioja – one of the smallest regions in Spain – is known for its wine.

Spain's main industries are shown on map **I**. Over the past thirty years tourism has grown into the biggest industry of all. It is concentrated along the Mediterranean coast.

(I) Spain's main industries

KEY

Important industries

🚗 Cars	⊙ Engineering	🟨 Tourism
🧪 Chemicals	Metal	Intensive farming (fruit and salad)
💡 Electrical	Shipbuilding	Wine
🍅 Food processing	✂ Clothes and shoes	🐟 Fishing
	€ Banking and finance	

activity...

1 Look at this pie chart. It shows the proportion of people working in primary, secondary and tertiary industries in Spain.

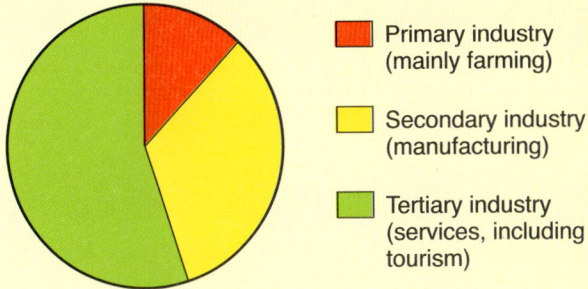

- ■ Primary industry (mainly farming)
- ■ Secondary industry (manufacturing)
- ■ Tertiary industry (services, including tourism)

a) Make a large copy of the chart. List each of the industries shown on map **I** in the correct sector of the chart.

b) Estimate the percentage of the workforce in each sector.

aim high...

2 You are going to compare two regions of Spain: Catalonia and Andalucia.

a) Look at map **I**. List the types of industry in each region.

b) Draw two pie charts to show employment in the two regions. Use the figures in the table below.

Region	Primary	Secondary	Tertiary
Catalonia	2.7	37.0	60.3
Andalucia	9.6	24.8	65.6

c) Write a few sentences to compare industry and employment in the two regions.

d) Catalonia has always been a richer region than Andalucia. Suggest why.

e) Over the past 30 years Andalucia has been catching Catalonia up. Suggest why.

In Catalonia Catalan is spoken, not Spanish. Six million people live in Catalonia with their own flag, government and even football team (Barcelona!).

holiday plans...

Investigate one region in Spain as a holiday destination. You could choose the region where you want to do an activity, or the region where your chosen city is (if they are in different regions, you could investigate both).

Answer these questions for each region:

What is the landscape like? What activities can you do? (use pages 104 to 105)

What is the climate like? When is the best time to go? (use pages 106 to 107)

Which cities are there? What attractions do they have? (use pages 108 to 109)

Will there be lots of tourists? What else could you find there? (use map **I**, page 110)

→ Spain in Europe

Spain, like the UK, is part of the EUROPEAN UNION (EU).
Originally formed in 1957, the EU is a group of European
countries that got together to TRADE more with each other.
The UK joined in 1973, and Spain in 1986. In recent years
the EU has expanded further. Spain is a good example of
the benefits that EU membership can bring.

J Europe and the European Union

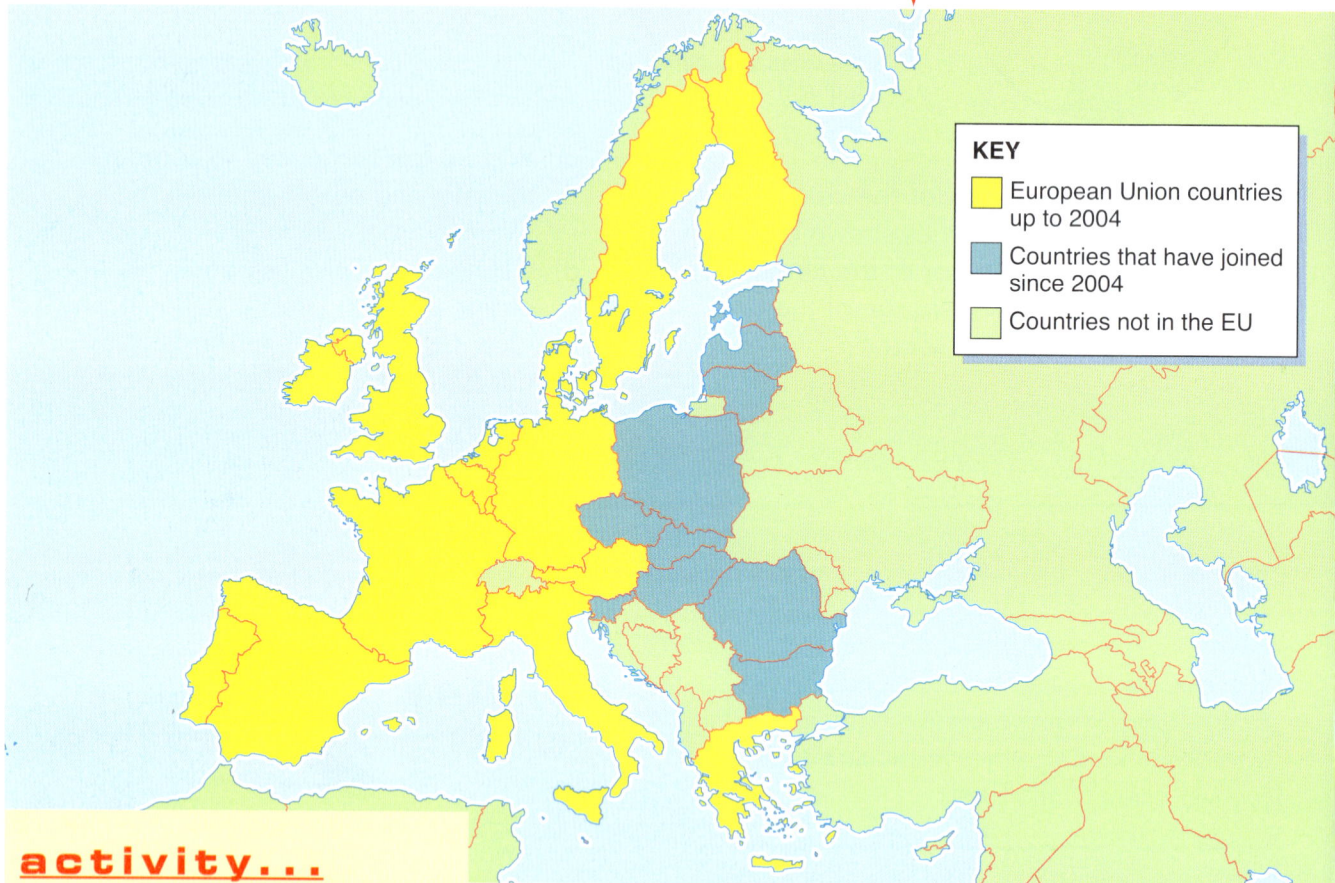

KEY

☐ European Union countries up to 2004

☐ Countries that have joined since 2004

☐ Countries not in the EU

activity...

1 a) Look at map **J**. How many European Union countries can you name? Label them on a copy of the map. Then, check page 135.

b) Name the European countries that are not in the European Union (not many!).

2 Look at the charts in **K**.

a) What percentage of Spain's i) exports, and ii) imports, are with EU countries?

b) Explain how joining the EU has helped Spain.

K Spain's exports and imports

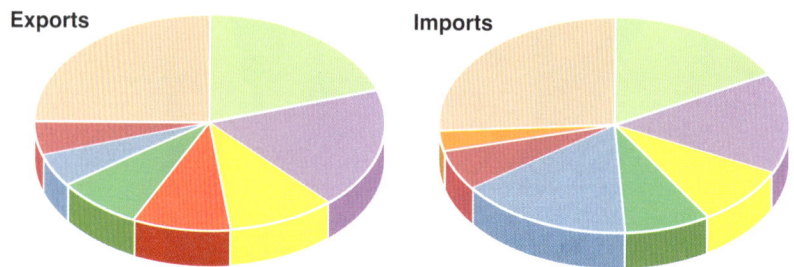

Exports   Imports

KEY

☐ France
☐ Germany
☐ Italy
☐ Portugal
☐ UK
☐ US
☐ Japan
☐ Other EU countries
☐ Other countries

When Spain joined in 1986 it was one of the poorest countries in the EU …

Today, Spain is no longer poor. It is quickly catching up its European neighbours.

aim high...

3 You work for the government in a country that is thinking about joining the EU. They want to persuade ordinary people that it would be a good thing for their country. Write a TV advert to give the reasons for joining the EU. Use Spain's experience to help you to make your case.

holiday plans...

When you go to Spain you will need to use Euros. The exchange rate for pounds (£) and Euros is roughly £1 = 1.3 Euros (check the latest rate).

You are going to Spain for a week. How much money will you take? Think about the things you need money for (accommodation, food, getting around and things you plan to do).

→Tourism takes off…

So far, you have travelled all around Spain and hardly been near a beach! That's about to change. TOURISM is now a major industry in Spain and an important part of the Spanish economy. So, today, you are going to the beach to find out more about tourism.

Benidorm is on the Costa Blanca on the Mediterranean coast. Back in the 1960s, it was a sleepy fishing village (photo **L**). Today, you would hardly recognise it as the same place (photo **M**). Whatever happened to Benidorm?

L Benidorm in 1963

M Benidorm today

activity…

1 Compare photos **L** and **M**. Write a paragraph to compare Benidorm in the 1950s with the town today. Think about:
   - the size of the town
   - the number and size of buildings
   - what people do there.
2 Look at graph **N**. You are going to turn it into a living graph.
   a) Make a copy of the graph on a page in your book. Allow space to continue the graph to 2020. (It will help to turn the page on its side.)
   b) Read the events below. Decide when each event would have most likely happened. Label each one at the correct point on the graph:
   - Peter and Rosie go back to Benidorm with their children. There are lots of British tourists.
   - Pablo sells his hotel to an international company and retires to his villa.
   - Pablo gives up fishing and opens a hotel.
   - Peter and Rosie go to Benidorm for a quiet honeymoon.
   c) What will happen to the number of British tourists in Benidorm in the future? Make a prediction and continue your graph up to 2020. Make up one more event on the graph.

(1) Back in the 1950s, a few smart businessmen came to Benidorm and thought it would make a good tourist **RESORT**.

(2) So they rented hotel rooms, booked airline flights and advertised for people to come. The **PACKAGE HOLIDAY** was born.

(3) News got around. The Costa Blanca became very popular. Tourism turned Benidorm into a high-rise resort.

(4) Most people either love it or hate it. Some tourists now look for smaller, quieter resorts. So what is the future for Benidorm?

Number of tourists

1950   1960   1970   1980   1990   2000

Number of tourists visiting Benidorm

SPAIN

Benidorm

Costa Blanca

holiday plans...

The good thing about a package holiday is that someone else makes all the plans. Unlucky, you have got to make your own plans! Again, the Spanish tourism website will help at www.spain.info.

a)   Click on 'Advanced search'. A list of options should appear.

b)   Select a region, province, island or city from the options. Type it in the box.

c)   Then, tick one, or more, of the search categories e.g. youth hostels, beaches, airports.

d)   Click 'Search'. A list of links should appear. You should find all the information you need to plan your holiday. You should plan: how to get there, where to stay, what to do and when to go.

115

➜ The perfect holiday

Not long now until your holiday. Just time to do a bit of last minute shopping and pack your bags. But, first, let's go through your holiday plans …

■ your final task…

You need all the right ingredients for a perfect holiday – the right place, an easy journey, exciting activities, comfortable accommodation and (hopefully!) good weather. One small detail missing from your plans – like booking the hotel – and the holiday could go badly wrong.

It's rather like preparing the perfect paella. Forget one vital ingredient, and you could have some very unhappy customers in your restaurant!

You are going to bring together all the holiday plans that you have made in this unit for a perfect holiday. To make sure that nothing is left to chance you are going to write your plans down. Think of it a bit like a recipe. You could write your plans like this …

RECIPE FOR A PERFECT HOLIDAY

1 Where to go …

2 How to get there …

3 What to do …

4 When to go …

5 Where to stay …

6 What to bring …

Where to go

Are you going to spend the holiday in one place, or travel around? Which region, or regions, will you go to? Will you visit a city?

	APRIL		
MO		4	11
TU		5	12
WE		6	13
TH		7	14
FR	1	8	15
SA	2	9	16
SU	3	10	17

How to get there

What means of transport will you use? Is there an airport, a station or a port nearby? If you are moving around on your holiday, how will you travel?

Where to stay

What accommodation are you going to stay in? Which one have you chosen? Where is it?

What to do
What activities will you do? Which attractions will you visit?

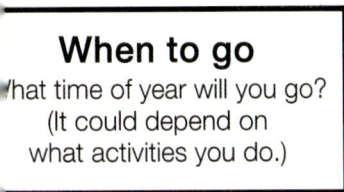

When to go
What time of year will you go? (It could depend on what activities you do.)

Paella is a mixture of seafood, vegetables, chorizo (spicy sausage) and rice, cooked together with olive oil in a large pot – all the ingredients come from Spain.

What to bring
Finally, you need to pack your bags. What do you need to take? How much money do you need? Make a list of things to pack in your suitcase. (It will depend on the time of year and the activities you hope to do.)

▌▌ coming up...

We all need energy. In some parts of the world people walk miles to collect firewood so that they can cook a meal. Here, we simply flick a switch to turn on the light.

▌▌ through the unit.

You will look at energy through the eyes of two teenagers living in different parts of the world. You will find out how they use energy, the problems with burning fuels, and what the alternatives are.

■ your final task...

You will re-design each other's homes to make them more energy-efficient.

starter...

1 Look at the image of the Earth at night.
   a) Identify five areas where people use lots of energy, e.g. the USA
   b) Identify five areas where people use little energy, e.g. Northern Canada
You can use the world map and an atlas to help you.

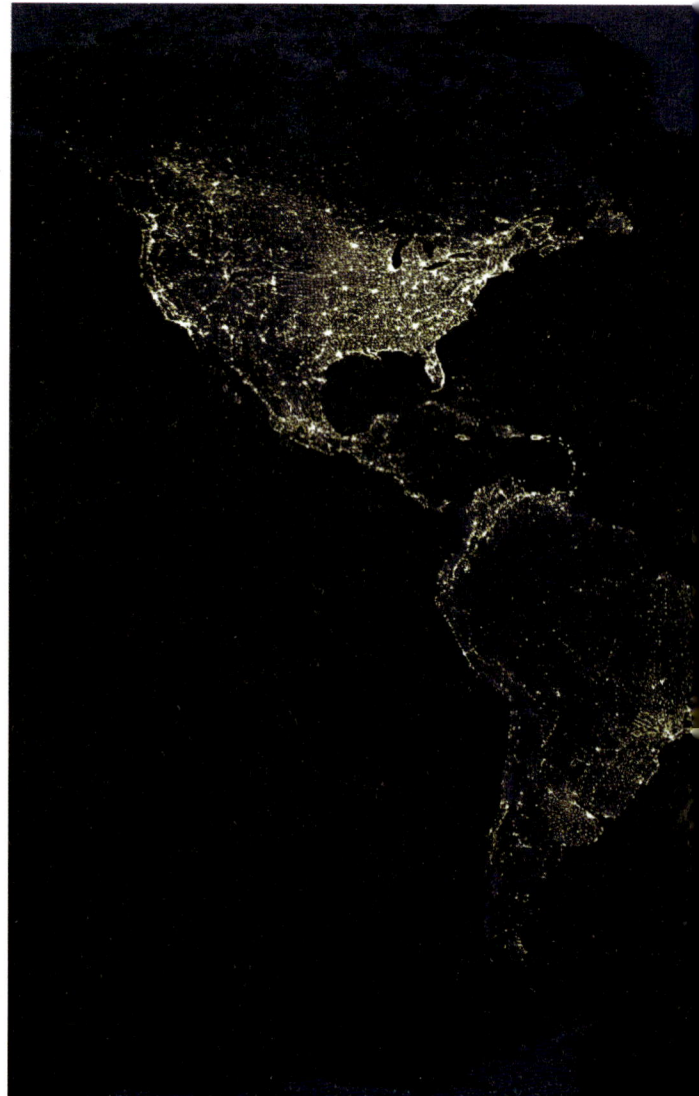

UK →

Kenya

Energy audit...

Through this unit you are going to work with a partner. One of you will become an expert on the UK, the other one will become an expert on Kenya. (Don't worry if you don't know much about Kenya yet. You will be given all the information you need.)

First, decide who is going to become the expert for each country. Each of you will produce an energy audit for your country. Use a whole page in your book for this. As you go through the unit, you will answer questions about energy in your country. You will use your answers at the end of the unit.

Start on this page with the question: How much energy does your country use? Find the UK or Kenya on the satellite image in order to answer the question.

Here is an image of the Earth at night, seen from space. The lights show where people live and are using energy – the brightest spots of light show cities. The image gives us clues about the amount of energy used around the world.

→ We all need energy

Mark lives in a modern house in the UK. Almost everything he touches seems to use energy. It's probably the same in your home.

Once upon a time, the energy in our homes came from burning FUELS like wood or coal. These days most of our energy comes in the form of electricity. You might think electricity is a clean form of energy, but don't be fooled! Most of our electricity is still produced by burning fuel.

activity...

1 Look at drawing **A**, or think about your own home.
   a) List all the things that use energy in, and around, the home, e.g. *washing machine*.
   b) Tick the things that use electricity.
   c) For things that don't use electricity, name the fuels that they use.

A Inside Mark's house in the UK

activity...

2 Read Angela's description of the way her family uses energy.

  a) List the things they do that you would use electricity for, e.g. washing clothes.

  b) What problems does lack of energy cause the family? Mention three things.

aim high...

3 Imagine what it would be like to live without electricity. Write a story to describe a day without electricity. You could call it, The day the lights went out.

Angela's family live in a small village in Kenya. Like most villages in rural Kenya, there is no electricity.

Energy is something that people think about every day in my village. Each morning I get up at 6.00 am. Before we go to school, my sister and I go searching for firewood. Two bundles of wood are just enough for cooking each day. Everyone in the village needs wood, so trees get chopped down and the forest gets smaller. As the years go by, we walk further and further to find wood.

While we are at school in the morning, my mother and grandmother do the household chores – washing the clothes, sweeping the house, grinding the millet (our main food). Everything takes much longer if you haven't got electricity.

When we get home, dinner is cooking over the fire. We only light the fire twice a day – once in the morning and again in the evening. We also burn animal dung to make the wood last longer. The problem with burning fuel is the smoke. We have to do all our cooking outside, even in the rainy season.

In the evening, I study by the dim glow of the paraffin lamp. Without it we would have no light at all. But, for poor families like ours, paraffin is expensive.

energy audit...

Make notes to answer this question for your country: What sources of energy do people use in their homes?

B Angela's home in Kenya

→ Energy everywhere

There are many sources of energy (drawing **C**). They fall into two main types.

- NON-RENEWABLE RESOURCES can be used only once when they are burnt to provide energy.
- RENEWABLE RESOURCES can be used over and over again without being used up.

C Energy sources

The **SUN** is where the Earth gets all its energy from.

WIND turns turbines on land or offshore.

WAVES have energy that can be harnessed by wave machines.

NUCLEAR FUEL, like uranium, is radioactive (it breaks down to give energy). It is used in nuclear power stations.

TIDAL power is captured by a barrier built across the river mouth.

OIL and **GAS** are drilled from rocks beneath the land or sea.

Pipes carry oil and gas onshore where they are refined and stored.

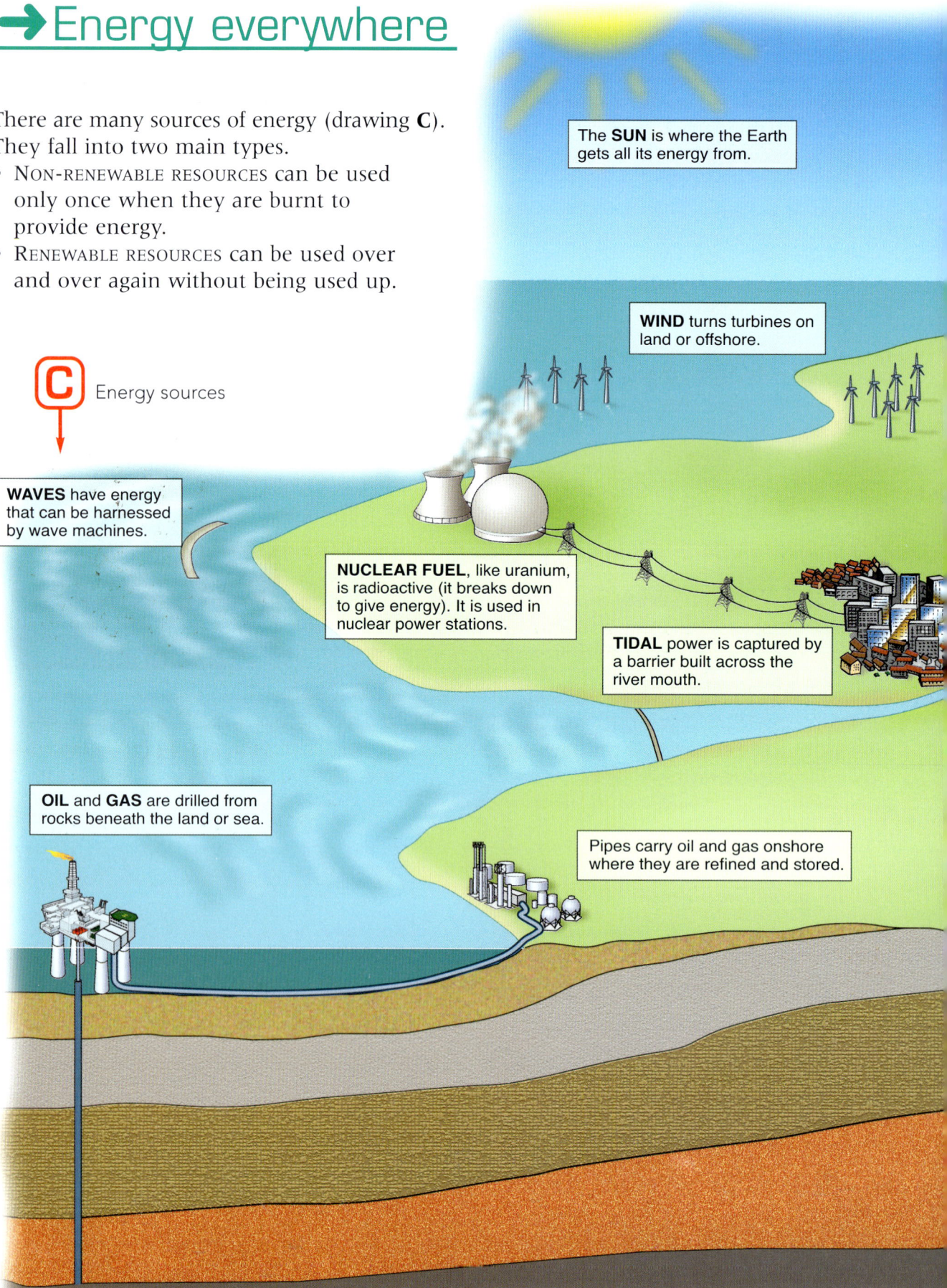

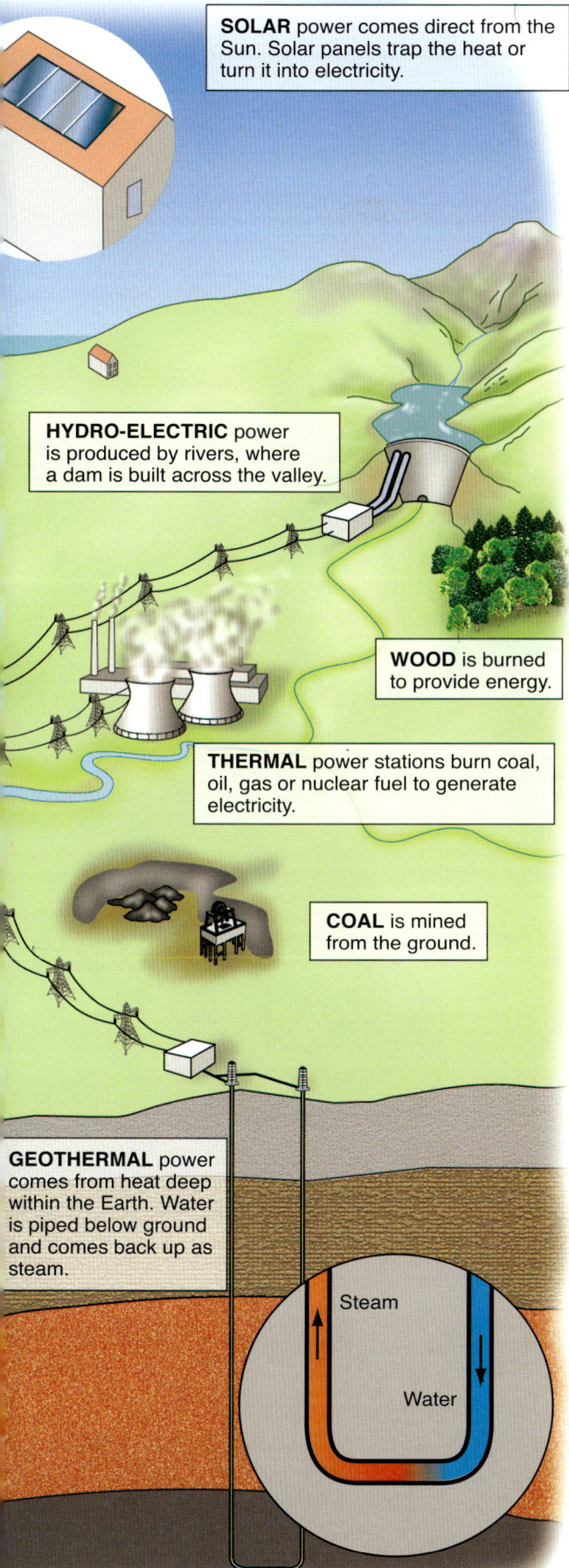

SOLAR power comes direct from the Sun. Solar panels trap the heat or turn it into electricity.

HYDRO-ELECTRIC power is produced by rivers, where a dam is built across the valley.

WOOD is burned to provide energy.

THERMAL power stations burn coal, oil, gas or nuclear fuel to generate electricity.

COAL is mined from the ground.

GEOTHERMAL power comes from heat deep within the Earth. Water is piped below ground and comes back up as steam.

Steam

Water

activity...

1 Look at drawing **C**. Sort the energy resources into two groups: Non-renewable and Renewable.
2 Read these newspaper headlines.

> SOIL EROSION DANGER AS FORESTS DISAPPEAR

> PIPELINE LEAK THREATENS COASTAL WILDLIFE

> RESIDENTS BLAME HEALTH PROBLEMS ON RADIATION

> VILLAGES FLOODED FOR NEW POWER SCHEME

a) Match each headline to one of the energy resources in drawing **C**.
b) In each case, explain the link between the headline and the energy resource.
c) Which energy resource do you think is likely to do:
   i) least damage to the environment?
   ii) most damage to the environment?
   Give reasons.

energy audit...

Do some research in an atlas. Make notes to answer this question for your country:
What energy resources does your country have?
a) Does the country have a coastline, mountains or rivers where energy could be obtained? Use a *physical* map.
b) Does the country have high temperatures or high rainfall? Use a *climate* map.
c) Does the country have any fossil fuels? Use a *resource* map.

➜ From fuel to fumes

Burning fuels causes air pollution. In the UK, most of our pollution comes from POWER STATIONS and traffic. Every time we switch on the light or go somewhere in the car we are adding to the problem. People in Kenya use less energy than we do, but they have pollution problems too!

D How Mark and Angela obtain energy

Power stations burn thousands of tonnes of fuel every day. It heats water and turns it into steam.

Steam hits the blades of a turbine and makes it spin. The turbine turns the generator to produce electricity.

Angela walks for miles to collect wood and bring it home.

Tall chimneys emit waste gases like carbon dioxide (the main cause of global warming) and tiny particles of soot and ash. The wind carries them away and they end up somewhere else.

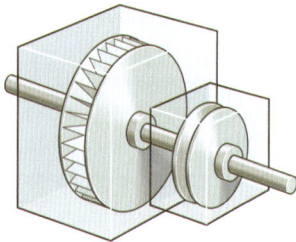

At a coal-fired power station, steam cools back to water inside large cooling towers. 60% of the energy produced goes into the air!

Wood is burnt on an open fire. Smoke gets everywhere.

Cables carry the electricity to cities. Each large power station can produce enough electricity for a whole city.

Electricity is used in our homes for everything from cooking to computers.

Heat is used for cooking.

Nearly half the world's population still cook with solid fuel (like wood, charcoal or coal) in their home. The smoke from burning these fuels turns kitchens into death traps. Particles in the smoke are the cause of many health problems, including lung disease, cancer, pneumonia and asthma. In Kenya, 96 per cent of people don't have electricity in their homes. Wood and dung provide most of the energy that is used.

Around the world smoke in the home from cooking on wood kills nearly one million children a year. It is a bigger killer than malaria, and is the fourth main cause of death in poor countries, like Kenya.

Cooking on a wood fire in a Kenyan home   **E**

activity...

1 Look at source **D**.
Here is a simple flow diagram to show how people in Kenya obtain energy from wood.

collect wood ⟶ burn wood ⟶ smoke
⟶ heat for cooking

Now, draw your own flow diagram to show how people in the UK obtain energy from burning coal in a power station. Use the information in source **D** to help you. The arrows below will get you started. You can draw cartoons or write labels in the spaces.

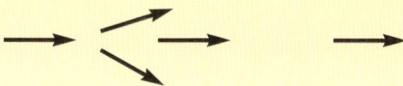

2 Look at the table on the right comparing the UK and Kenya.
   a) Which country i) is most wealthy ii) uses most energy iii) creates most CO_2?
   b) What is the connection between wealth and CO_2 emissions?

aim high...

3 Asthma is a problem for many people in Kenya and the UK. Recent studies have shown that it is linked with air pollution.
   a) Kenya is a poor country with little air pollution, yet many people suffer from asthma. Why is this?
   b) The number of children in the UK with asthma is increasing. Suggest why.
   c) Suggest how the risk of getting asthma could be reduced in i) Kenya ii) the UK.

energy audit...

Make notes to answer this question for your country: *What pollution problems does your country have?*

	UK	Kenya
Population (millions)	61	37
Wealth per person ($US)	26,240	1,010
Energy consumption (kg oil equivalent per person)	3,982	500
Carbon dioxide emissions (tonnes per person)	9.6	0.3

➔ Global warming

The biggest problem of all, from burning fossil fuels, is GLOBAL WARMING.

The atmosphere around the Earth acts like a natural greenhouse. Carbon dioxide (CO_2), and other gases in the air, traps some of the Sun's heat, allowing the rest to escape. This is called the GREENHOUSE EFFECT (drawing **G**). Without it the Earth would either boil or freeze. By burning fossil fuels we are producing more CO_2. As a result more heat is trapped and the temperature is rising.

G The Greenhouse Effect on a small scale

The Sun's rays pass through the glass and warm the greenhouse

Some heat escapes through the glass. If it didn't the greenhouse would get too hot

Some heat is trapped by the glass. This keeps the greenhouse warmer than outside

I grow nicely in here

I'd wilt if it got any hotter

H Changes in global temperature since 1860

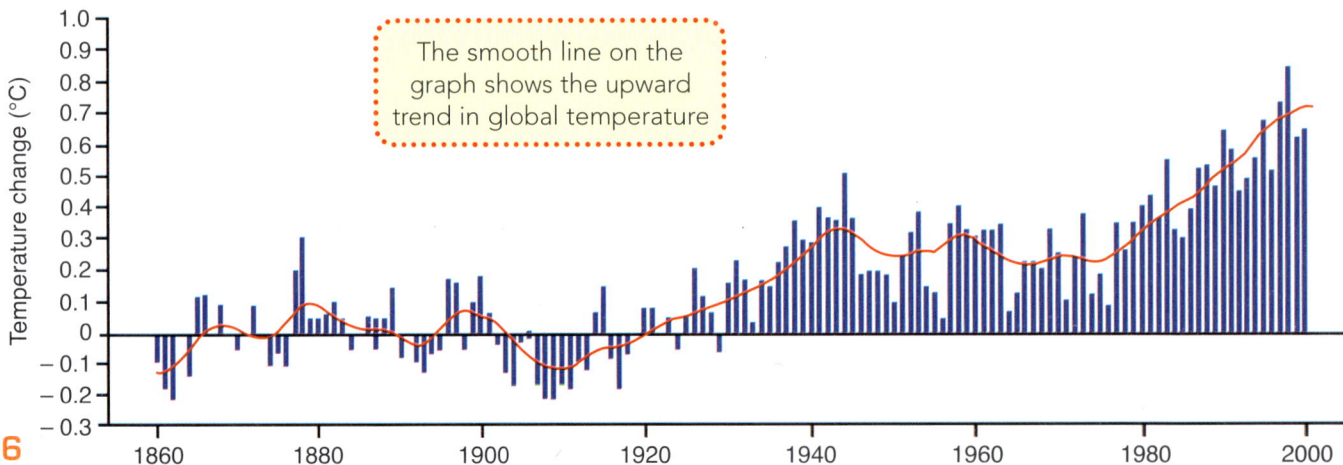

The smooth line on the graph shows the upward trend in global temperature

Temperature change (°C)

Graph **H** shows that throughout the twentieth century the Earth was warming up. Most of the hottest years have been recorded since 1990. Scientists predict it will get much hotter over the next hundred years.

So, what's the big problem? We all like hot weather, don't we? Well, here are a few things to think about.

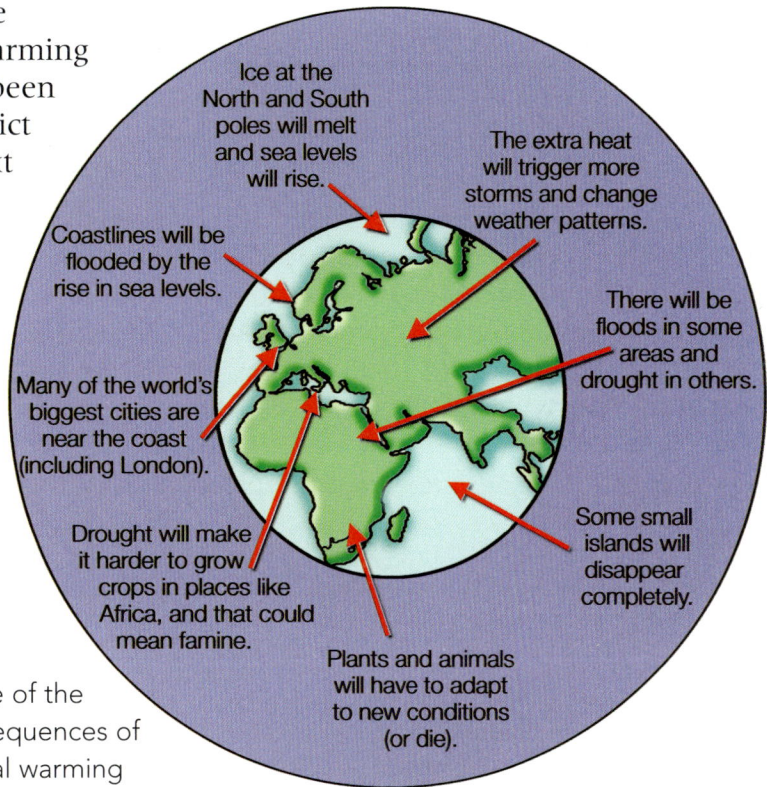

Ice at the North and South poles will melt and sea levels will rise.

The extra heat will trigger more storms and change weather patterns.

Coastlines will be flooded by the rise in sea levels.

There will be floods in some areas and drought in others.

Many of the world's biggest cities are near the coast (including London).

Drought will make it harder to grow crops in places like Africa, and that could mean famine.

Some small islands will disappear completely.

Plants and animals will have to adapt to new conditions (or die).

I Some of the consequences of global warming

activity...

1 Look at drawing **G**. It shows the greenhouse effect on a small scale.
   a) Make a large drawing like this to show the greenhouse effect on a global scale.
   b) Label the 'Earth' and the 'Atmosphere' on your drawing.
   c) Label the arrows using the labels on drawing **G**. Replace the words 'greenhouse' and 'glass' with 'Earth' and 'atmosphere'.

2 Look at graph **H**.
   a) Name the coldest decade (ten-year period) and the warmest decade.
   b) Carefully, describe how global temperature has changed since 1860. Here are some words you can use:

   > trend   average   increase   warming
   > exceptional   fluctuation   global

   c) From what the graph shows, how would you expect global temperature to change in your lifetime? Explain your answer.

discuss...

3 Talk with a partner.
   a) How do you feel about global warming? Is it something that worries you or something you just accept? Why?
   b) If it worries you, what could you do about it?

energy audit...

Make notes to answer this question: *How could your country be affected by global warming?* Source **I** will give you some ideas. You could also do some research using this website: www.climatehotmap.org
   a) Click on a region – either Europe or Africa. A map will appear.
   b) Click on a number on the map for more information.

➜ Free as the wind

So far ...

You have looked at the way things are: how people use energy, where it comes from and the problems that this can cause. Above all, you have learned that burning fossil fuels leads to global warming. But what can be done? It's time to look at alternatives.

Study pages 128 and 129 if you are the UK expert

Wind has so much energy that it gets everywhere! For centuries people have used wind powcr for sailing ships and turning windmills. Now, we can turn the wind's energy into electricity. Wind is free, and it will never run out (it's renewable). It sounds like the perfect source or energy – or is it?

J A WIND FARM in Cornwall. Wind farms are groups of wind turbines that produce electricity. They are like mini-power stations.

Wind is free, but the turbines are expensive. Overall, wind power is slightly cheaper than electricity from coal.

Some people think that wind farms spoil the landscape. Other people think they are attractive. It's a matter of opinion really.

Turbines only take up 1 per cent of the land on a wind farm. Farmers can still use the land for grazing animals.

Birds sometimes fly into the blades on the turbines and are killed.

The rotating blades on the turbine make a wooshing noise. This can disturb people living nearby.

Wind is unreliable. Some days it can be windy, but on other days it is calm. Wind turbines generate electricity for about 80 per cent of the time

activity...

1 Do a simple cost-benefit analysis to decide whether to build a new wind farm. Cost-benefit analysis is a method for weighing up the good points (benefits) and bad points (costs) before making a decision about a new development.

a) Look at the factors around photo **J**. How important is each factor in the decision to build a wind farm? Rank the factors from 1 to 8. Give a score of 8 to the most important factor, 7 to the next most important, and so on.

b) Sort the factors into two groups: costs and benefits. Some will depend on your opinion. Write them in a table like this, with their score.

Costs	Score	Benefits	Score

c) Add the scores. Write the total scores for the costs and benefits at the bottom of the table.

d) Compare the total scores. Would you decide to build a wind farm, or not? Write a paragraph to explain your decision.

aim high...

2 You have been asked to advise the government on the best way to produce energy in the future.

a) Think about the costs and benefits of building a new coal power station, like the one on page 124. Do a simple cost-benefit analysis for a power station.

b) Using your analysis, compare a coal power station with a wind farm. Would you recommend building more power stations or more wind farms? Write a short report to justify your choice.

energy audit...

Make notes to answer this question for the UK: *How useful is wind power as an alternative source of energy? Are there other alternatives?* (Look back at pages 122 to 123 for ideas.)

Wind turbines take up a lot of space. This wind farm covers about one km^2. It generates enough power for a small town. Imagine how much space a large city would require.

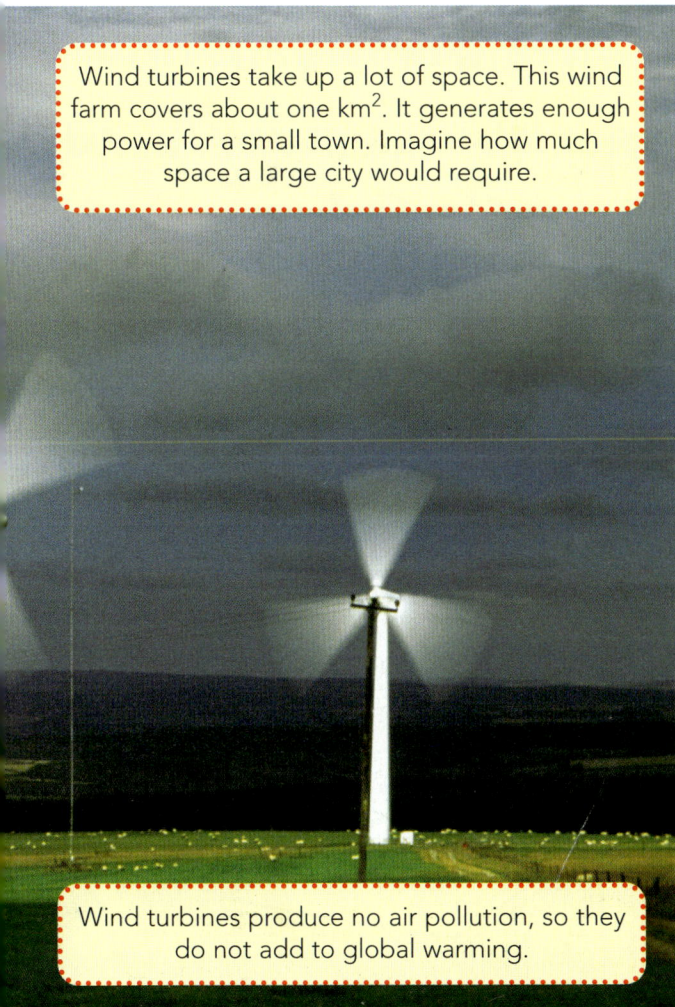

Wind turbines produce no air pollution, so they do not add to global warming.

Wind farms in the UK in 2007. More are being built each year. The government is aiming for 10 per cent of our energy to come from renewable sources by 2010.

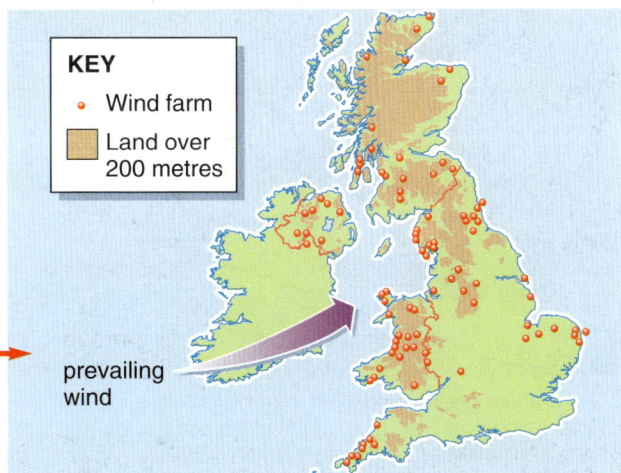

K

KEY
- Wind farm
- Land over 200 metres

prevailing wind

129

➔ Small-scale solutions

Study pages 130 and 131 if you are the Kenya expert

Life can be hard in Kenya. A typical family in rural Kenya spends hours collecting firewood – time that could be spent on childcare, education or earning money. They can also spend one-third of their income on fuel, such as paraffin for lighting.

As the population of Kenya grows, more land is needed for farming, and forests disappear. Eventually, there won't be any trees left to burn. Then what? The people in one village have found an alternative – a hydropower project.

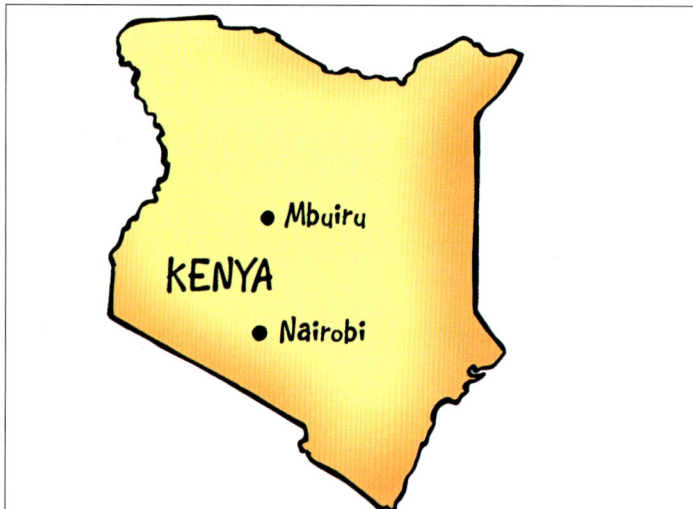

L

Mbuiru is a small village in Kenya, 200 km north of the capital, Nairobi. People are poor and the area suffers from drought. How could energy help to solve their problems?

At first sight, water seems an unlikely source of energy. Drought is so bad that some of the rivers sometimes run dry. But the River Tugbu, near Mbuiru, flows all year round.

The villagers got involved. First, they had to dig a canal to divert the water in the river.Then they had to make a water storage tank and install the turbines to generate the electricity.

Two years later everything was ready. The villagers held their breath, and then ... power!

The hydropower project in Mbuiru has many benefits:

- people have light in their homes
- people pay less for energy than before
- children are better educated
- people no longer need to burn wood
- people have more money
- people have more time
- people can use electrical appliances (like sewing machines)
- people are more healthy
- fewer trees are chopped down
- there is less smoke in homes
- people run small businesses
- the environment improves.

M

Help–we're drowning!

...gers met to discuss a small-scale hydropower project. Many ...e unsure. They had heard about big dams that had flooded ...ges. However, they decided to go ahead with the project.

...this is how it works. Water is diverted from the river. It flows ...g the canal. The water is stored in a tank. Then it flows down ...es under the force of gravity. Water drives the turbines to ...erate electricity – enough to light ninety homes in the village.

activity...

Make a concept map to show the benefits of the hydropower project in Mbuiru.

a) Draw a lightbulb in the middle of a page in your book.

b) Make copies of the benefits listed in **M** on small pieces of paper (or your teacher will give you copies to cut out).

c) Place the benefits on your page around the hydropower project and stick them down. Draw lines to show the direct links between the benefits and the hydropower project. For example, *People have light in their homes*.

d) Now, find links between the benefits and draw more lines. Annotate the lines to explain the links. For example, *There is less smoke in homes **because** people no longer need to burn wood.*

At the end your page could look like this:

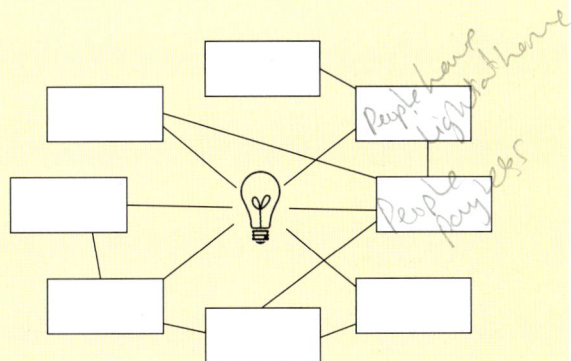

energy audit...

Make notes to answer this question for Kenya: *How useful is hydropower as an alternative source of energy? Are there other alternatives?* (Look back at pages 122 to 123 for ideas.)

→ Sustainable lifestyles

The way that we use energy is not SUSTAINABLE. In the UK, we depend on fossil fuels that one day will run out. In Kenya people burn wood, but the trees are disappearing. All over the world, people are worried about the effects of global warming.

The answer – for the UK and Kenya – is to find more sustainable lifestyles that use renewable energy and don't harm the environment. But it's easier said than done!

In Kenya, we don't have enough energy to meet our basic needs. Where can we find the energy?

In the UK, we still depend on fossil fuels and use more energy than we really need. How can we change our habits?

■ your final task...

1 Through this unit you have become an expert in one country: Kenya or the UK. Now you are going to share the information with your partner. (You will need this information to re-design each other's homes in activity 2.)

Ask your partner these questions to find out more about their lifestyle:
- how much energy does your country use?
- what sources of energy do people use in their homes?
- what energy resources does your country have?
- what pollution problems does your country have?
- how could your country be affected by global warming?
- how useful could alternative sources of energy be?

2 Re-design your partner's home to make their lifestyle more sustainable.
a) First, look back at your partner's home on page 120 or 121. Think about the information they gave you in activity 1. Is their lifestyle sustainable? If not, how could you make it more sustainable?
b) Look at some alternatives on the page opposite. Which ideas could you use in your design?
c) Draw a design for your partner's home. You could completely re-build it, or just change it a bit. It's up to you. Explain how your design would make it more sustainable.
d) Show your design to your partner and they will show you their design for your home. Evaluate your partner's design. Would you like to live there? Is it more sustainable?

Large windows

Large windows facing the Sun help to trap heat (like a greenhouse). Double or triple glazing will keep the heat in at night.

Biogas converter

A biogas converter stores animal dung and converts it into methane gas. It can be burnt to provide heat.

Solar panel

Solar panels on the roof can be used to heat water or to generate electricity.

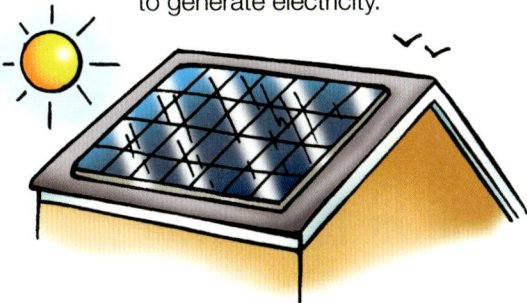

Clay oven

Clay ovens are more efficient than an open fire. They only use half of the fuel.

Smoke hood

Smoke hoods take smoke directly from the fire up a chimney and out of the home.

Soil on roof

Soil on the roof is a good insulator (does not let heat through). It keeps the building cool in summer and warm in winter.

Shower

Showers use less water than a bath - and so they use less heat too.

Low-energy light bulb

Low-energy light bulbs use only 1/5 energy of an ordinary bulb. They last longer too.

Wind turbine

A single wind turbine can produce enough electricity for a group of homes (but only when the wind blows!).

→ Ordnance Survey 1:25,000 map symbols

Customer Information

Whilst we have endeavoured to ensure that the information in this product is accurate, we cannot guarantee that it is free from errors and omissions, in particular in relation to information sourced from third parties.

Reproduction in whole or part by any means is prohibited without the prior written permission of Ordnance Survey.

Ordnance Survey, the OS Symbol, OS and Explorer are registered trademarks of Ordnance Survey, the national mapping agency of Great Britain.

Made, printed and published by Ordnance Survey, Southampton, United Kingdom.

General information about the work of Ordnance Survey, its products and services is available from Customer Service Centre, Ordnance Survey, Romsey Road, SOUTHAMPTON, United Kingdom, SO16 4GU.

Phone: Customer Service Centre 08456 05 05 05 (Lo-call).   Fax: 023 8079 2615.

Web site: www.ordnancesurvey.co.uk/leisure

Communications

ROADS AND PATHS   Not necessarily rights of way

Motorway   Service area   7 Junction number
M1 or A6(M)
Dual carriageway
A35
Main road
A30
Secondary road
B3074
Narrow road with passing places
Road under construction
Road generally more than 4 m wide
Road generally less than 4 m wide
Other road, drive or track, fenced and unfenced
Gradient: steeper than 20% (1 in 5); 14% (1 in 7) to 20% (1 in 5)
Ferry; Ferry P - passenger only
Path

RAILWAYS

Multiple track | standard
Single track | gauge
Narrow gauge or
Light rapid transit system (LRTS) and station
Road over; road under; level crossing
Cutting; tunnel; embankment
Station, open to passengers; siding

PUBLIC RIGHTS OF WAY   (Rights of way are not shown on maps of Scotland)

Footpath
Bridleway
Byway open to all traffic
Restricted byway

Restricted byway (from 2nd May 2006 roads used as public paths were redesignated as restricted byways. They provide a right of way for walkers, horse riders, cyclists and other non-mechanically propelled vehicles).

Public rights of way shown on this map have been taken from local authority definitive maps and later amendments.

Rights of way are liable to change and may not be clearly defined on the ground.
Please check with the relevant local authority for the latest information.

The representation on this map of any other road, track or path is no evidence of the existence of a right of way.

OTHER PUBLIC ACCESS

Other routes with public access (not normally shown in urban areas)

The exact nature of the rights on these routes and the existence of any restrictions may be checked with the local highway authority. Alignments are based on the best information available.

National Trail / Long Distance Route; Recreational Route

Permissive footpath   Footpaths and bridleways along which landowners have permitted public use but which are not rights of way.
Permissive bridleway   The agreement may be withdrawn.

Traffic-free
cycle route

National cycle network
route number - on road
1 National cycle network
route number - traffic free

Scotland

In Scotland, everyone has access rights in law over most land and inland water, provided access is exercised responsibly. This includes walking, cycling, horse-riding and water access, for recreational and educational purposes, and for crossing land or water.

Access rights do not apply to motorised activities, hunting, shooting or fishing, nor if your dog is not under proper control. The Scottish Outdoor Access Code is the reference point for responsible behaviour, and can be obtained at www.outdooraccess-scotland.com or by phoning your local Scottish Natural Heritage office. "Land Reform (Scotland) Act 2003

National Trust for Scotland,
always open - limited opening - observe local signs

Forestry Commission Land / Woodland Trust Land

England & Scotland

Firing and test ranges in the area. Danger! Observe warning notices
Champs de tir et d'essai. Danger! Se conformer aux avertissements
Schiess- und Erprobungsgelände. Gefahr! Warnschilder beachten
Visit www.access.mod.uk for information.

DANGER
AREA

ACCESS LAND

Access land boundary and tint
Access land in woodland area
Access information point

England

Portrayal of access land on this map is intended as a guide to land which is normally available for access on foot, for example access land created under the Countryside and Rights of Way Act 2000, and land managed by the National Trust, Forestry Commission and Woodland Trust. Access for other activities may also exist.
Some restrictions will apply; some land will be excluded from open access rights. The depiction of rights of access does not imply or express any warranty as to its accuracy or completeness. Observe local signs and follow the Countryside Code.
Visit www.countrysideaccess.gov.uk for up-to-date information

MANAGED   Access permitted within managed
ACCESS   controls for example, local byelaws
           Visit www.access.mod.uk
           for information

General Information

VEGETATION   Limits of vegetation are defined by positioning of symbols

Coniferous trees
Non-coniferous trees
Coppice
Scrub
Bracken, heath or rough grassland
Marsh, reeds or saltings
Orchard

GENERAL FEATURES

+ Place of worship
  with tower
  with spire, minaret or dome
Current or former
place of worship
Building; important building
Glasshouse
Youth hostel
Bunkhouse/camping barn/other hostel
Bus or coach station
Lighthouse; disused lighthouse; beacon
Triangulation pillar; mast
Windmill, with or without sails
Wind pump; wind turbine
Electricity transmission line
Slopes
pylon pole

BP/BS Boundary post/stone
CG   Cattle grid
CH   Clubhouse
FB   Footbridge
MP; MS Milepost; milestone
Mon   Monument
PO   Post office
Pol Sta Police station
Sch   School
TH   Town hall
NTL   Normal tidal limit
W; Spr Well; spring

Gravel pit
Other pit or quarry
Sand pit
Landfill site or slag/spoil heap

BOUNDARIES

National
County (England)
Unitary Authority (UA), Metropolitan District (Met Dist), London Borough (LB) or District (Scotland & Wales are solely Unitary Authorities)
Civil Parish (CP) (England) or Community (C) (Wales)
National Park boundary

HEIGHTS AND NATURAL FEATURES

52 Ground survey height
284 Air survey height

Surface heights are to the nearest metre above mean sea level. Where two heights are shown, the first height is to the base of the triangulation pillar and the second (in brackets) to the highest natural point of the hill.

Vertical face/cliff
Contours may be at 5 or 10 metres vertical interval

Loose rock   Boulders   Outcrop   Scree
Water   Mud   Sand; sand & shingle

ARCHAEOLOGICAL AND HISTORICAL INFORMATION

+ 1066   Site of antiquity   VILLA Roman
×   Site of battle (with date)   Castle Castle
         Visible earthwork   Non-Roman

Information provided by English Heritage for England and the Royal Commissions on the Ancient and Historical Monuments for Scotland and Wales

Selected Tourist and Leisure Information

RENSEIGNEMENTS TOURISME ET LOISIRS SÉLECTIONNÉS   AUSGEWÄHLTE INFORMATIONEN ZU TOURISTIK UND FREIZEITGESTALTUNG

P&R Parking / Park & Ride, all year/seasonal
     Parking / Parking et navette, ouvert toute l'année/en saison
     Parkplatz / Park & Ride, ganzjährig/saisonal

i   Information centre, all year /seasonal
     Office de tourisme, ouvert toute l'année/en saison
     Informationsbüro, ganzjährig/saisonal

V   Visitor centre
     Centre pour visiteurs
     Besucherzentrum

     Forestry Commission visitor centre
     Commission Forestière Centre de visiteurs
     Staatsforst Besucherzentrum

PC   Public convenience
     Toilettes
     Öffentliche Toilette

Telephone, public/roadside assistance/emergency
Téléphone, public/borne d'appel d'urgence/urgence
Telefon, öffentlich/Notrufsäule/Notruf

Camp site /caravan site
Terrain de camping /Terrain pour caravanes
Campingplatz /Wohnwagenplatz

Recreation/leisure /sports centre
Centre de détente /loisirs /sports
Erholungs-/Freizeit-/Sportzentrum

Golf course or links
Terrain de golf
Golfplatz

Theme/pleasure park
Parc à thèmes/Parc d'agrément
Vergnügungs-/Freizeitpark

Preserved railway
Chemin de fer touristique
Museumseisenbahn

Walks/trails
Promenades
Wanderwege

Cycle trail
Piste cyclable
Radfahrweg

Mountain bike trail
Chemin pour VTT
Mountainbike-Strecke

Horse riding
Équitation
Reitstall

Public house/s
Pub/s
Gaststätte/n

Viewpoint
Point de vue
Aussichtspunkt

Picnic site
Emplacement de pique-nique
Picknickplatz

Country park
Parc naturel
Landschaftspark

Garden/arboretum
Jardin/Arboretum
Garten/Baumgarten

Nature reserve
Réserve naturelle
Naturschutzgebiet

Fishing
Pêche
Angeln

Water activities
Jeux aquatiques
Wassersport

Slipway
Cale
Helling

Other tourist feature
Autre site intéressant
Sonstige Sehenswürdigkeit

Cathedral/Abbey
Cathédrale/Abbaye
Kathedrale/Abtei

Museum
Musée
Museum

Castle/fort
Château/Fortification
Burg/Festung

Building of historic interest
Bâtiment d'intérêt historique
Historisches Gebäude

National Trust

English Heritage

Historic Scotland

➜ Reference map of Europe

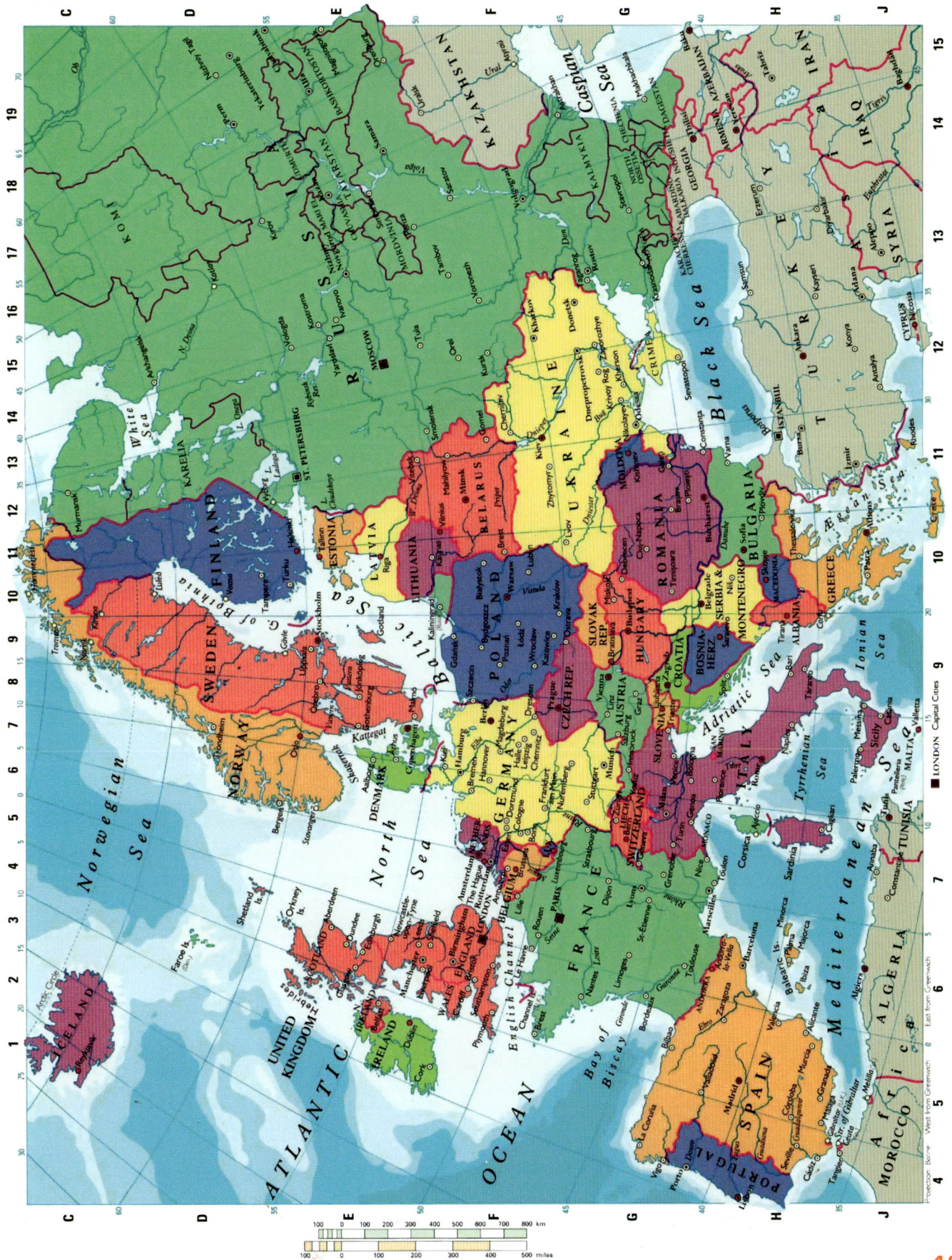

Coverage of Key Concepts

■ Main Key Concept   ▨ Other Key Concepts

Book 2	Place	Space	Scale	Interdependence	Environmental interaction and sustainable development	Physical and human processes	Cultural understanding and diversity
1 Welcome to Earth Village	Imaginary village – representative of the world	Global population distribution – exemplified by Earth Village	Interplay between local (village) and global	Impact of population change + resource consumption by rich + poor countries	Optimistic + pessimistic views of human impact on planet	Population growth, birth + death rates	Values and attitudes to family size + resource consumption
2 Landscape detective	Yorkshire Dales	Location of limestone landscapes in England	Regional/national		Urban landscape as a reflection of local geology	Rock cycle, weathering + erosion, limestone formation	
3 Here is the weather forecast	UK and western Europe	Weather patterns in th UK. Satellite image + weather chart for western Europe	National/continental			Cloud and rain formation. Passage of a depression	
4 Once upon a coal mine	Easington Colliery, north-east England	Distribution of former coal mining areas in UK. Changing industry in NE	Zooming in + out between personal, local, regional, national + global	Loss of old industry and arrival of new industries in NE		Social + economic decline	Work expectations + factors that influence them
5 A question of football	Wimbledon/Milton Keynes Ivory Coast/Europe	UK transport network. Location of cities + population distribution in Europe	Enquiries at a range of scales from local to global	Relocation of a football club. Migration from Africa to Europe			Attitudes to women's football + foreign players
6 Coast to coast	Four coastal locations in the UK – Porlock Bay, Marloes, Happisburgh, Winchelsea	Choosing the best location for a new coastal resort on 1:25,000 OS maps	Local/national		Potential impact of a new resort on the coastal environment	Coastal processes – erosion, deposition, longshore drift	
7 Spain – here we come!	Spain – physical + human characteristics. Pupils' imaginations of Spain	Location of features in Spain. Distribution of climates, population + economic activites	National, including regional variation	Tourism to Spain. Economic development within the EU		Development of a tourist resort.	Development of pupils' imaginations of Spain. British tourism in Spain
8 Changing fumes	UK/Kenya	Pattern of global energy consumption. Distribution of energy sources in the UK	Personal, local, national + global	Consequences of our energy use in other places	Impact of different energy sources on environment. Sustainable energy use	Greenhouse effect/global warming	Exchange of ideas + experiences of sustainable living in UK + Kenya

→ Glossary

A

AIR PRESSURE   the weight of the air pressing down on the Earth's surface

ANTICYCLONE   an area of high pressure that brings fine dry weather

ARCH   an opening formed when waves wear right through a rocky headland

ATMOSPHERE   the layer of air around the Earth

ATTRITION   how rocks and stones get worn away by bumping into each other

B

BEDDING PLANE   the line between two layers of sedimentary rock

BIRTH RATE   the number of births per thousand people

C

CALL CENTRE   a large base for services provided over the phone (like banking or insurance)

CAVE   a hollow space in rock formed by erosion (caves are common in limestone where water dissolves the rock)

CLIFF   a very steep rocky slope

CLOUD   a mass of water droplets or ice crystals in the atmosphere

COAL   a sedimentary rock made of dead plant material, it is used as a fuel

COAL SEAM   a layer of coal found within another sedimentary rock

COALFIELD   an area where coal can be mined

COLD FRONT   the line where cold air pushes sharply under warm air giving heavy rain

CONVECTIONAL RAINFALL   rain caused by warm air rising when the ground is heated

CORRASION   how a cliff is eroded by rocks carried by waves, acting like sandpaper

D

DEATH RATE   the number of deaths per thousand people

DEPOSIT   to drop material (for example, waves deposit sand to form a beach)

DEPRESSION   a moving area of low pressure, which usually brings wet windy weather

DRY VALLEY   a valley without a river at the bottom (characteristic of limestone areas)

E

ECOLOGICAL FOOTPRINT   the amount of space needed to produce all the resources we use and to get rid of our waste

EMPLOYMENT   paid work

ERODE   to wear away (for example, waves erode the base of a cliff)

EUROPEAN UNION   an organisation that links 27 countries in Europe (encouraging trade and migration)

F

FAULT   a crack in the Earth's crust where the rocks have moved

FETCH   the distance over the sea across which the wind blows

FOSSIL   remains of a plant or animal preserved in rock

FOSSIL FUEL   coal, oil or gas formed from the remains of plants or animals found in rock

FRONT   the dividing line between two air masses

FRONTAL RAINFALL   rain caused by warm air rising over cold air at a front

FUEL   something we use to provide energy (usually by burning it)

G

GEOLOGY   the study of rocks

GEOTHERMAL POWER   energy obtained from heat deep in the Earth's crust

GLOBAL WARMING   rising temperatures around the world

GLOBALISATION   the way jobs, people and ideas move around the world

GREENHOUSE EFFECT   the natural way gases in the atmosphere keep the air warmer

GROYNE   a barrier built down a beach to slow movement of material along the coast

H

HYDRAULIC ACTION   how water pressure breaks up rock at the Earth's surface

HYDRO-ELECTRIC POWER   electricity generated by fast flowing water

I

IGNEOUS ROCK   rock formed when molten rock cools and solidifies

INFANT MORTALITY   the number of deaths of children under one year old for every thousand born

ISOBAR   a line on a weather map joining places with equal air pressure

J

JOINT   a vertical crack in a rock

L

LANDSLIDE a sudden movement of rock, soil and vegetation down a slope

LIFE EXPECTANCY the number of years a person can expect to live (on average)

LIMESTONE PAVEMENT a horizontal outcrop of limestone which looks like paving slabs

LONGSHORE DRIFT the movement of material along the coast, carried by waves

M

MANUFACTURING INDUSTRY making products from raw materials in a factory

METAMORPHIC ROCK rock that has been changed by the effects of heat or pressure

N

NON-RENEWABLE RESOURCE a resource that when it is used up is gone forever

P

PACKAGE HOLIDAY a holiday where everything is organised for you by a travel company

PLATEAU an area of high, flat land

POPULATION the number of people in a place

POPULATION DISTRIBUTION the way people are spread within an area

POPULATION EXPLOSION the dramatic growth of the world's population

POPULATION GROWTH RATE the difference between birth rate and death rate

POTHOLE a vertical hole in a limestone area where a joint has been enlarge by solution

POWER STATION a place where electricity is generated

R

RECLAIMED LAND an area that was once sea and has been drained to create dry land

REGION an area with a number of distinctive characteristics (for example, a mountain region)

RELIEF RAINFALL rain caused by air being forced over hills or mountains

RENEWABLE RESOURCE a resource that will never run out

RESOURCE something that we use to survive (like food or fuel)

ROCK CYCLE the continual formation and destruction of rocks in the Earth's crust over millions of years

S

SCAR a steep outcrop of bare rock

SEDIMENT rock fragments deposited by water

SEDIMENTARY ROCK rock formed by depositing sediment (mainly on the seabed)

SERVICE a type of industry that helps (or serves) people

SOLAR POWER energy from the Sun

SOLUTION how certain rocks are dissolved by water (for example, limestone)

SPIT a strip of sand or shingle in the sea

STACK a pillar of rock standing in the sea

STALACTITE an icicle-shaped deposit hanging from the roof of a limestone cave

STALAGMITE a column of calcium carbonate growing from the floor of a limestone cave

STRATA layers of sedimentary rock

SUSTAINABLE something that can continue because it is not wasteful

SWALLOW HOLE a pothole (in a limestone area) down which a river flows

T

TOURISM an industry specialising in holidays

TRADE the buying and selling of goods between countries

TRANSNATIONAL COMPANY a large company with branches in more than one country

TRANSPORT to carry material (for example, sand or shingle carried by waves)

U

UNEMPLOYMENT lack of employment or work

W

WARM FRONT a line where warm air rises over cold air, giving prolonged rain

WAVE-CUT NOTCH a dent worn away at the base of a cliff by wave erosion

WAVE-CUT PLATFORM a flat rocky outcrop left behind when waves erode a cliff away

WIND FARM a group of wind turbines that generate electricity from wind

→ Index

→ Acknowledgements

The Publishers would like to thank the following for permission to reproduce copyright material:

Photo credits
Cover l © Eye Ubiquitous / Alamy, r Robert Harding Picture Library; **pp. 6–7** BARBARA SAX / AFP / Getty Images; **p. 14** tl © Craig Lovell / CORBIS, tr Moodboard / Alamy, bl © Tibor Bognar / Corbis, br © Redlink / Corbis; **pp. 20–21** © Manor Photography / Alamy; **p. 20** t Goss Images / Alamy, b Corbis RF / Alamy; **p. 21** t © Made and Found / Alamy, b © Fotosonline / Alamy; **p. 22** t © Jim Sugar / CORBIS, b Ern Mainka / Alamy; **p. 23** Paul Springett / Alamy; **p. 24** Trevor Smithers ARPS / Alamy; **p. 28** Michael Sayles / Alamy; **pp. 30–31** John Widdowson; **p. 36** © Gareth Fuller / PA Photos; **p. 41** © NEODAAS / University of Dundee; **p. 42** Peter Hannert / Photonica / Getty Images; **p. 43** James Osmond / Alamy; **p. 47** © NEODAAS / University of Dundee; **p. 48** © NEODAAS / University of Dundee; **p. 50** © Universal / Everett / Rex Features; **p. 51** Bob Anderson / Rex Features; **p. 52** t John Widdowson, c © Photographer's Choice / Getty Images, b Tetra Images / Getty Images; **p. 55** John Widdowson; **p. 56** John Widdowson; **p. 57** John Widdowson; **p. 58** © AirFotos / Construction Photography; **p. 59** © Photographer's Choice / Getty Images; **p. 60** © www.webbaviation.co.uk; **p. 62** Tetra Images / Getty Images; **pp. 64–65** Tom Purslow / Manchester United via Getty Images; **p. 66** David Rogers / Getty Images; **p. 67** © Stuart Roy Clarke (www.homesoffootball.co.uk); **p. 70** Ian Walton / Getty Images; **p. 71** Martin Rose / Bongarts / Getty Images; **p. 72** London Aerial Photo Library; **p. 73** t © London Aerial Photo Library, b David Rogers / Getty Images; **p. 76** t Clive Mason / Getty Images, b © Ben Radford / CORBIS; **p. 80** Camera Lucida / Alamy; p. 81 t © Mike Page, bl © Andrew Stacey, br © London Aerial Photo Library; **p. 84** © Andrew Stacey; **pp. 90–91** © Chris Warren / Loop / CORBIS; **p. 93** t © Mike Page, b Robert Estall photo agency / Alamy; **p. 94** © Mike Page; **p. 98** © Andrew Stacey; **p. 100** tl AzureRepublicPhotography / Alamy, tr © Martin Child / The Image Bank / Getty Images, b Mark Eveleigh / Alamy; **p. 101** t Alex Segre / Alamy, bl Nicholas Stubbs / Alamy, br Denis Doyle / Getty Images; **p. 102** Glow Images / Alamy; **p. 105** t Chris Knapton / Alamy, tml Mark ZYLBER / Alamy, tmr FAN travelstock / Alamy, bml © INTERFOTO Pressebildagentur / Alamy, bmr Picture Contact / Alamy, b Nature Picture Library / Alamy, br © Fancy / Veer / Corbis; **p. 107** l Cro Magnon / Alamy, r © Alex Segre / Alamy; **p. 108** Bernhard Lang / Digital Vision / Getty Images; **p. 109** t Kevin Schafer / The Image Bank / Getty Images, c Kevin Foy / Alamy, b Denis Doyle / Getty Images; **p. 111** © LLUIS GENE / AFP / Getty Images; **p. 114** t Phillip / Fox Photos / Getty Images, b JOSE JORDAN / AFP / Getty Images; **p. 119** NASA Goddard Space Flight Center (NASA-GSFC); **p. 120** © Randy Faris / Corbis; **p. 121** John Widdowson; **p. 125** ADAM HART-DAVIS / Science Photo Library; **pp. 128–129** John Martin / Alamy; **p. 132** l John Widdowson, r © Randy Faris / Corbis.

t = top, b = bottom, l = left, r = right, c = centre

Text credits
p. 10–11 US Census Bureau, population pyramid, International Database; **p. 16** United Nations, Population Division/DESA, graph of population growth; **p. 61** 'At your service: Indians heed call of the West' by Charles Haviland, The Guardian (October 18, 2003); **p. 71** percentage of women amongst footballers, from The First World Atlas of Football by Radovan Jelinek and Jiri Tomes (Bohemia Hobbies, 2002); 'Mexican football club signs woman' by Jo Tuckman, The Guardian (December 18, 2004), adapted; **p. 77** News International Syndication, 'An African footballer's story' by Kenneth Akpueze, The Times (2007); **p. 94** Eastern Daily Press, 'Coastal erosion leaves house on the edge' by Maria Fulcher, Eastern Daily Press (December 23, 2003); **p. 102** Office for National Statistics, bar graph of most popular holiday destinations for UK tourists, reproduced with the permission of the Controller of OPSI, © Crown copyright.

Maps credits
pp. 32, 82–83, 87, 91–92, 96 Ordnance Survey maps, reproduced by permission of Ordnance Survey on behalf of HMSO, © Crown copyright; **p. 134** OS map symbols © Crown copyright 2008; **p. 135** Map of Europe, taken from *Philip's Modern School Atlas* (2006), reproduced by kind permission of Octopus Publishing Group.

Every effort has been made to contact and acknowledge ownership of copyright. The publishers will be glad to make suitable arrangements with any copyright holders whom it has not been possible to contact.